Leaders in American Geography

Leaders in American Geography

Volume 1: Geographic Education

Thomas F. Barton, Indiana University

and

P.P. Karan, University of Kentucky

NEW MEXICO GEOGRAPHICAL SOCIETY
Mesilla, New Mexico

New Mexico Geographical Society
2916 Snow Road
Mesilla, New Mexico 88046

This book is printed on acid free paper.
First Edition, 1992

Typeset, design and layout by Gyula Pauer.

Published in association with
PRESTIGE BOOKS INTERNATIONAL

Contents

Part 2 The Synthesizers

Part 3 The Professionals

List of Plates

New Mexico Geographical Society

The objective of the Society is to foster the discovery and dissemination of geographical knowledge. Associated with the Society is the Center for American Places of Harrisonburg, Virginia.

Preface

This book is part of a two-volume series dealing with leaders in American Geography. This first volume focuses on leaders in American geographic education. The second volume will focus on leaders in geographic research.

The idea for the book was conceived during a conversation with Professor Thomas F. Barton in 1977. In the early 1980s Professor Barton and I sought the names of individuals who would be included in this first volume. We asked one hundred prominent geographers to submit the names of ten geographers whom they felt were leaders in geographic education in America. From these, we selected only those names which were on the lists of at least five respondents.

These leaders laid the foundation of geography in high schools, colleges and universities in the United States. They trained most of the geographers who are teaching in our schools and colleges today. Most of the leaders have been recognized by the Association of American Geographers and the National Council for Geographic Education for their contributions to geography and geographic education.

This collection of biographies spans the entire history of American geography. All individuals repesented here share or have shared the common bond of love of education. When the untapped potential of a student meets the liberating art of a teacher, learning begins. A special relationship is born. These leaders built these special relationships. Such encounters defy description, but they leave an indelible imprint.

These biographies are based on materials gathered from various sources, including personal correspondence with geographers around the country. I wrote Barton's biography. The biographies of Clarence F. Jones and J.B. Jackson were prepared by Cotton Mather. Despite my best efforts, I was unable to obtain photographs and the precise dates of birth and death for several persons included in this book, and one person considered his date of birth private information which need not be made public!

Professor Barton died in August 1985 while this book was nearing completion. I have organized and edited the materials prepared by Professor Barton for this volume. I also wrote the biographies of most of the leaders included in Part 3.

P. P. Karan

June 1, 1992

Acknowledgments

I am grateful to many persons who contributed their ideas and recollections of the individuals included in this book. I have relied on Mamie L. Anderzhon, Thomas F. Barton, Clarence F. Jones, Preston E. James, Clyde F. Kohn, Cotton Mather, Joseph R. Schwendeman, C. Langdon White, and Joe Russell Whitaker for their memories of many people in this book.

Tsering Tsomo ably put most of the manuscript on the word processor. I would like to express my special gratitude to Hazel Karan for her rigorous and sensitive editing of the entire text. My thanks go as well to Gyula Pauer for making superb photographic prints for reproduction. I am indebted in particular to three anonymous reviewers for their helpful suggestions.

I would like to express my appreciation to the Editorial Advisory Board of the New Mexico Geographical Society and the Center for American Places for their support and enthusiasm for the project.

We would like to thank the following for permission to reproduce the photographs:

University of Northern Iowa, Special Collections Library, Cedar Falls, Iowa for the photograph of Professor Alison Aitchison.

Clark University Archives for the photographs of Professors Wallace Atwood, Douglas Ridgley, and Samuel Van Valkenburg.

The University of Chicago, Joseph Regenstein Library for the photographs of Alice Foster, Charles Colby, Robert Platt, and J. Paul Goode.

J. Baylor Roberts, National Geographic Society, for the photograph of Gilbert H. Grosvenor.

Eastern Washington University, Archives and Special Collections, for the photograph of Otis W. Freeman.

National Council for Geographic Education and James W. Vining, for the photographs of Katheryne Thomas Whittemore, Herbert Henry Gross, Phillip Bacon, Elizabeth Eiselen, Henry Warman, and Alfred Herman Ludwig Meyer.

The George Arents Research Library at Syracuse University for the photograph of George B. Cressey.

Anne McMaster, University of Toronto for the photograph of Griffith Taylor.

Sophia Shaw, Leesburg, Florida for the photograph of Earl B. Shaw.

Mrs. John W. Morris, Norman, Oklahoma for the photograph of John W. Morris.

J. Russell Whitaker, Nashville, Tennessee for the photographs of Almon E. Parkins, and Ray Hughes Whitbeck.

Bert E. Burns, Mankato, Minnesota for the photograph of Cora P. Sletton.

University of Pittsburgh for the photographs of Zoe Thralls and Mary Viola Phillips.

Low Memorial Library, Columbiana Collection, Columbia University for the photographs of J. Russell Smith, Armin K. Lobeck,and Richard E. Dodge.

Indiana University Archives for the photographs of George J. Miller and Stephen S. Visher.

The John Hopkins University Archives for the photograph of Isaiah Bowman.

Harvard University Archives for the photographs of William Morris Davis and Nathaniel Southgate Shaler.

Carroll College, Waukesha, Wisconsin for the photograph of Benjamin F. Richason.

Mamie Anderzohn, Shenandoah, Iowa for the photograph of Edith Parker.

University of Wisconsin, Madison for the photograph of Vernor C. Finch.

Cotton Mather for the photograph of John Brickerhoff Jackson.

Eastern Michigan University Archives for the photograph of Mark Jefferson.

James J. Parsons for the photograph of Carl Sauer.

University of Kentucky Archives for the photograph of Ellen Churchill Semple.

Geoffrey J. Martin, Easton, Connecticut for the photograph of Olive E. Baker.

Erselia Barton, Bloomington, Indiana for the photograph of Thomas F. Barton.

Jackie Kalmes, Northwestern University for the photograph of Edward Espenshade.

Thomas P. Grimes for the photograph of Richard Hartshorne.

Valley City State College, Valley City, North Dakota for the photograph of Ina Cullum Robertson.

The Library of Congress for the photograph of John Wolter.

The George Arents Research Library at Syracuse University for the photograph of George B. Cressey.

Anne McMaster [illegible] for the photograph of [illegible] Taylor.

[illegible] for the photograph of [illegible]

[illegible] for the photograph of [illegible] W. [illegible]

[illegible] for the photographs of [illegible] D. Barrows, and [illegible] Wheeler.

[illegible] for the photograph of [illegible]

University of Chicago for the photographs of [illegible] Phillips.

[illegible] Library, Columbia University for the photographs of [illegible] E. [illegible]

Indiana University Archives for the photographs of George [illegible] and Stephen S. Visher.

The Johns Hopkins University Archives for the photograph of [illegible] Bowman.

Harvard University Archives for the photographs of William [illegible] Davis and [illegible]

Smith College [illegible] for the photograph of [illegible]

[illegible] University [illegible] for the photograph of [illegible] Parkins.

University of Wisconsin [illegible] for the photograph of [illegible]

[illegible] for the photograph of [illegible] Jackson.

Introduction to Leaders in American Geographic Education

We are at a critical juncture in American geography. Over the years geographers have been active leaders and participants in a seemingly endless series of problems and issues involving violence and conflict at home and abroad, environmental deterioration, scarcity of energy and other resources, rapid rates of urban growth and development, racism, and other situations. For nearly a century geographers have shown a deep interest and concern in social, economic, and political problems. Not only has this been true of past generations of geographers, but it is equally characteristic of present-day geographers. A growing number of geographers are becoming involved in the study of collective social behavior, social processes, race and gender issues, social organization and social institutions, and in the making of public policy to improve socioeconomic conditions.

We are devoting a lot of time to recalling the contributions of the past generation of geographers, and remembering who we are and from where we have come. There is a desire to make new commitments. The importance of recalling the past is that it opens the imagination to things both old and new, and sets us to thinking differently about the future. Biographies personalize this historical experience. These biographies will give us a better understanding of ourselves through the discovery of a shared experience, a similar training of the intellect, an idea affirmed, a disappointment repeated. The objective of reading these leaders' biographies is to restore memories, to review our commitments, and to extend our sense of camaraderie.

The life and times of the leaders in geographic education noted in this book span a period of more than one hundred and fifty years. The book concentrates mostly on the first half of the twentieth century, roughly from the beginning of modern geography in America at the turn of the century to the end of the Second World War. This richly textured period represents a special experience in American geography.

Before the turn of the century there was no visible organized geography in the United States. Between 1890 and 1910 geography emerged with a few key individuals at its center - Nathaniel S. Shaler and William Morris Davis at Harvard; J. Russell Smith at the University of Pennsylvania; Richard E. Dodge at Teachers College, Columbia University; and Rollin D. Salisbury and Ellen Churchill Semple at Chicago. Mark Jefferson was at Michigan State Normal College, and R. H. Whitbeck was supervising state normal and model schools in New Jersey. The American Geographical Society of New York, having completed its first half century, was examining its policy and developing areas of research in which it could accomplish the most for the American people. The American Association for the Advancement of Science had a section devoted to Geology and Geography, and in 1904 the Association of American Geographers was organized.

At the same time, educators were deploring the meager offerings and ineffective teaching methods of geography in our secondary schools. Geography had shared the widespread interest in science that characterized the latter half of the nineteenth century; physiography, recommended by the National Education Association (NEA) Committee of Ten, had gained ground as a high school subject during the 1890s (Rosen, 1947:409). Physical geography, with its genetic classification and interpretation of land forms, presented ideas relatively near the frontier of geographic research (Powell, 1911:119). However, in only a few schools was the subject taught well enough to arouse student enthusiasm, since few teachers had the training necessary to do more than teach from a textbook and there was little opportunity for field study.

In 1897 the *Journal of Geography* was established to meet the needs of a growing interest in geographic education. The *Journal* was later merged with the *Bulletin of the American Bureau of Geography*. Since 1902 the *Journal of Geography* has appeared under its present title.

Seven departments of geography were established in the United States between 1883 and 1906: at Princeton in 1883, at Cornell in 1894, at Columbia in 1896, at California in 1898, at Chicago in 1903, and at Nebraska in 1906. By 1912 twenty-four state universities offered geography courses (Winstead, 1912:312). Graduates of the University of Chicago Department of Geography played a major role in the development of geography. Many of the newer geography departments were staffed by geographers trained at Chicago. For example, Bernard Scheckel followed Charles Dryer at the State Normal School, Terre Haute, Indiana; V.C. Finch became an assistant at the University of Wisconsin; and Charles C. Colby became an instructor in geography at the State Normal School, Winona, Minnesota (Foster, 1923:3). By 1913 Carl O. Sauer and Stephen S. Visher were working on their doctorates at Chicago.

In the early twentieth century the time was ripe for an upsurge of activity in geography under strong leadership. Geographers such as William Morris Davis, Ellen Churchill Semple, R. H. Whitbeck, Almon E. Parkins, Douglas C. Ridgley, Rollin D. Salisbury, and J. Russell Smith provided leadership, generated ideas, and established new institutions. Their influence was felt as geography spread into other universities.

By 1920 students who had studied under these geographers were teaching throughout the country in universities, colleges, teacher-training institutions, and high schools. They also made important contributions to published research for the teaching of geography. Nathaniel S. Shaler (1841-1906) and his disciple, William Morris Davis (1850-1934) of Harvard, and Ellen Churchill Semple (1863-1932) at Chicago were giants among the early geographers. All were memorable teachers whose influence in physical and human geography is felt to this day.

The number of collegiate institutions which offered geography continued to grow in the 1920s. Based on a review of 398 American college catalogs, Ida Shrode found that 319, or 80 percent, offered geography courses. (Shrode, 1929: 189-201). Shrode noted that some work in geography was practically universal in the curricula of two-year normal schools and four-year teachers' colleges. In an essay on "The Place of Geography in Higher Education," Whitbeck reported on the growth in the number of universities which offered geography courses and the establishment of independent departments of geography (Whipple, 1933:41-49). Whitbeck stressed the development of graduate work and research in geography. The department of geography at Chicago was the only one to be established at a university whose president had declared its first and most important responsibility was research and the training of researchers. Generally, the soundness of this goal came to be recognized in other universities.

A major problem of early leaders was to find the direction in which geographic research and education should move. They experimented with the new problems of geographic education and tried to improve the status of geography at the high school level. However, by 1922 the percentage of high school students electing physical geography had dropped to 4.3 percent from 21.5 percent in 1905. Geography was also partially swallowed up by a survey course in history called social science. By the 1930s geography had almost diappeared from the high school curricula in the United States. In 1934, only 1.6 percent of high school students elected physical geography. Most liberal arts colleges no longer accepted geography courses for admission.

Harvard University removed physical geography from its admission list in 1936. The same year marked the publication of the last College Board examination in geography. Fairbanks (1927: 160) noted that "if geographers had been more united and more aggressive, the subject would not have suffered the eclipse in the schools." Dryer (1924:141) stressed that geography was crowded out by numerous new studies during the 1920s.

Despite major efforts by geographic educators such as Isaiah Bowman, George J. Miller, Mark Jefferson, Wallace W. Atwood, Gilbert H. Grosvenor, Edith P. Parker and Clarence F. Jones, no action was taken to improve the status of high school geography during the 1930s. But during the 1920s and 1930s geography textbooks authored by some of our leaders had a profound effect on college-level geographic education. For example, the textbook on economic geography coauthored by Clarence Jones was widely adopted as a college text. After 1940 Edith Putnam Parker began to offer graduate work in geographic education at the University of Chicago.

Following the end of the Second World War, the American educational system responded to the obvious need to improve goegraphic literacy. Due largely to the efforts of Charles C. Colby, J. Russell Whitaker, George Thomas Renner, Otis W. Freeman, Samuel Van Valkenburg, Thomas F. Barton, Preston E. James, Mamie L. Anderzhon and Clyde F. Kohn, geography, which had almost disappeared from the high school curricula by 1930, made a modest comeback. The percentage of high school students studying geography increased. However, the student enrollment in the junior high and middle schools dropped (Manson, 1981:244).

An event of steller significance to the development of geography at the high school level occurred in 1961 under the leadership of Gilbert White. The High School Geography Project was initiated to improve the content of courses in geography at the freshmen-sophomore level and to develop new instructional materials to use in these courses.

In 1984 the National Geographic Society, under President Gilbert M. Grosvenor, began to promote geography in the schools. Along with Gilbert M. Grosvenor, leaders in geographic education such as Charles F. Gritzner, Richard G. Boehm, Christopher L. Salter, Edward P. Hogan, and Anthony R. de Souza have emerged as the most influential leaders in improving geographic education. The National Geographic Society's education program includes a network of state-based geographic alliances which began in 1986 and has since spread rapidly throughout most of the United States. Geographic alliances work at the local level with classroom teachers, academic geographers, school administrators, and citizens working together to promote geographic education in their respective states. The National Geographic Society supports the summer institutes, workshops, and conferences which are organized by the alliances. Under the leadership of Barry Bishop, the Society offers a very valuable internship program for college students.

Following the Second World War, there was a major expansion of graduate programs in geography and a score of universities offered doctoral degrees. For instance, during 1952-57 Clark University granted 56 doctorates; Northwestern, 24; Chicago, 23; Washington, 21; Michigan, 18; Teachers College, Columbia University, 18; Syracuse, 18; California, 16; Ohio State, 16; Wisconsin, 14; Illinois, 14; Louisiana, 13; Maryland, 9; Nebraska, 8; UCLA, 7; Harvard, 7; John Hopkins, 7; Iowa, 6; Indiana, 5; Penn State, 4; Columbia, 3; Minnesota, 3; North Carolina,

3; and ten other universities, 12. The total was 305 degrees (Schwendeman, 1952 and 1956).

During the decades following the Second World War, geographers such as Preston James, Richard Hartshorne, Cotton Mather, Clyde F. Kohn, Robert S. Platt, and Olive E. Baker provided major leadership in the development of graduate-level studies and research in geography. Their leadership led to the development of a discipline devoted to the unified and comprehensive understanding of regions and landscapes. However, like the larger society from which it grew, geography contained numerous different emphases and approaches to the study of geographic phenomena.

The leaders presented here have been divided into three groups, which represent the three general phases of the development of geography and geographic education. Those in the first group are the *pioneers* of geography: Shaler, Semple, Davis, Salisbury, Bowman, Gilbert H. Grosvenor, and Huntington.

The second group of leaders are the great *synthesizers,* who integrated interests in environment, regions, culture, and economy. These synthesizers, who emerged in the 1930s and expanded their interests throughout the next three decades, include Colby, Cressey, Van Valkenburg, Renner, Visher, Whitaker, James, Mather, and Kohn.

The third group of leaders represent the *new professionals*. They span the interests of the early pioneers and the synthesizers in geographic education, but are concerned more with the analysis of particular situations and with the application of spatial theories and models. They are also committed to effective teaching at all levels of education, but especially at the undergraduate or collegiate level. They are dedicated to preparing future geographers who choose college or high school teaching as their career.

Since interesting people do not fit neatly into categories, these pioneers, synthesizers, and professionals may lose some of their clarity. Each leader in the three groups has made a difference, left an innovative mark in geography, and made a personal impact on the profession. Despite a diversity among these leaders, some common threads emerge. Their biographies shed light on the complex, often mysterious connection among individuals, institutions, and the environment. The leaders stirred ambition, excited admiration, and inspired fellow geographers, students, and the general public.

All of the leaders included here had the ability to see the logic in an emerging situation and to act on that insight. This is, in part, a cognitive ability, as one sees how one's goals may be reinforced by linking them to larger contexts and trends.

A second element of personality suggested by these leaders is their desire to "make a difference." They made a difference, not because they wished it, but because they willed it. Each was strongly motivated to place his or her individual imprint on the complex world of education and teaching. For some, such as Preston E. James and Clarence F. Jones, their personal style was low-keyed but assertive. Others, such as Rollin D. Salisbury and William Morris Davis, were more overly competitive, even combative, reminding us of individuals who have the will to conquer, the impulse to fight for their educational goals in geography. For all of the leaders, probably, the "will to conquer" was there; some sought only to conquer challenging academic and instructional problems, while others were motivated as well by the need to influence people inside and beyond their institutions.

In reviewing the lives of the men and women included here, one is struck by the importance of personal motivation in shaping organizational direction and the agenda, and by recurrent themes of "the American character," as described by Henry Steele Commanger (1950). For example, here we find the sustained, single-minded motivation and Puritan ethic embodied in George J. Miller, Charles C. Colby, and Edith P. Parker. In many of the other leaders, we find an irrepressible optimism — in the face of what others considered great obstacles — which was crucial to the actions of Davis, Bowman, Salisbury, Atwood, Platt, White, and most of the other leaders.

Each leader in this book has or had a strong will and knows or knew how to mobilize the will of others. Some leaders do tower over their contemporaries. But for all, the metaphors of height are peculiarly apt. They have to be able to see above the mundane and beyond the immediate in geographic education. They need that view from the mountain top. Leaders such as J. R. Whitaker, Gilbert M. Grosvenor, and John Wolter have that vision, and they deal with the concrete.

Some people live in the present, oblivious of the past and blind to the future. Some dwell in the past. A very few have the knack of applying the past to the present in ways that show them the future. Leaders in this book such Bowman, Sauer, Cressey, Mather, de Blij, Bishop, and de Souza have this knack.

Leadership must serve a purpose, and the higher the purpose the greater the potential stature of the leader. But purpose is not enough. A leader has to perform, produce results, and do so in a way that serves the higher purpose. Hogan, Grosvenor, de Souza, and Mather have demonstrated the ways in which geographic leadership is exercised for a higher purpose. They did not shrink from the battle; they entered the arena to promote geography.

Part 1 The Pioneers

Alison Esther Aitchison

Alison Esther Aitchison
October 1, 1872 - January 5, 1964

During the 1920s and 1930s Alison Aitchison was one of the most prominent geographers and the foremost leader of geographic education in Iowa. Her teaching methods were widely used by the many teachers who studied geography with her; they found her publications, lectures, curriculums, equipment, and classroom procedures invaluable. After nearly ten years of successful classroom teaching in Iowa's public schools, she taught for four decades at Iowa State Teachers College.

Through travel and field study, Aitchison learned geography first hand. She got a head start in travel when soon after her birth in Illinois her Scottish parents took her to their former home near Edinburgh, where she spent her childhood. Later, the family moved to Canada; they eventually settled in northwest Iowa.

After she began teaching, Aitchison visited Europe several times, always placing Scotland on her itinerary. Two of her reconnaissance trips were unique. Aitchison and a French-Spanish speaking friend were adventurous enough one summer in the 1920s to ship a Chevrolet to the continent so they could drive throughout Europe. She spent a strenous month in the Soviet Union with several Russian guides. During the summer of 1932, Aitchison joined a group of graduate students from the University of Chicago to do intensive field work in Britain. The Scandinavian and Central European countries, Southern Canada, the central plateau of Mexico, and most of the United States was also explored.

Rich experiences in the field, an excellent academic background, a strong belief in the value of geography, and a superior teaching talent contributed to Aitchison's success in geographic education. She began a systematic study of geography at Iowa State Normal University. Later, after transferring to the State University of Iowa, she specialized in geology and earned a bachelors degree in science. She was granted a masters degree from the University of Chicago in 1914.

Aitchison was an active member of the National Council of Geographic Education. Early in her career she joined a small group which laid the groundwork for the establishment of the National Council of Geography Teachers, the forerunner of the present organization — the National Council for Geographic Education (NCGE). After serving two years on the Executive Committee and on the Board of Directors, she became president of the National Council of Geography Teachers in 1936. She was honored with the National Council of Geography Teachers Distinguished Service Award in 1945.

Recognized for her high standards of scholarship and for her encourage-

Wallace W. Atwood

ment of graduates who continued their education in graduate schools, Alison Aitchison authored or coauthored numerous articles in state and national professional journals. Her first book, *Iowa State Geography*, was published in 1921. In the 1960s, in collaboration with a close associate, she prepared a grade school geography textbook, *Latin America, Africa and Australia*, which was part of the "Land and People of the World" series.

After she officially retired, Aitchison remained active. She continued traveling throughout Iowa, helping teachers implement the Social Science Curriculum, one of her greatest contributions to geographic education. Due to Aitchison's leadership, geography was made the core of this curriculum.

A colleague writes:

> While Miss Aitchison's professional contributions are notable, it is the modest, kindly, forthright, canny Scottish Iowan whom her many friends cherish and appreciate for wise counsel (Uttley, 1964:278).

Wallace Walter Atwood
October 1, 1872 - July 24, 1949

Wallace W. Atwood was President of Clark University for twenty-six years, founder and director of its Graduate School of Geography, and initiator of the magazine *Economic Geography* (1925). One of his graduate students and geography colleagues, Professor Henry J. Warman, wrote: "Dr. Atwood was a masterful teacher, an artist in the classroom, and a prolific writer" (Warman, 1949:350). By advice and example — first at Chicago, then Harvard, and later at Clark — he directed a host of students whose privilege it was to know and study with him.

A prominent American geographer, George B. Cressey, wrote that Atwood brought geography to more people than any other American professor. Cressey substantiated this statement by citing the following statistics: "(1) fifty million children gained their picture of the Earth through his textbooks, (2) thousands of university students were inspired as they listened to his classroom presentations, and (3) nearly a hundred completed their doctorate under his supervision" (Cressey, 1949:296-306).

President Atwood demonstrated an extraordinary capacity for work and accomplishment in a many-faceted life. As an author he wrote no less than thirty books, scores of articles, and produced numerous workbooks, manuals, exercises, and wall maps (Cressey, 1949:296).

Born in Chicago, Wallace Atwood received his college education at the University of Chicago. His interest in geology and geography dated from a visit to Devil's Lake, Wisconsin, with Professor Rollin D. Salisbury, during his sophomore year. After graduating in 1897, he became an Assistant Geologist for the New Jersey Geological Survey, and in 1898-99 served as a member of the Wisconsin National History Survey. Atwood returned to the University of Chicago as an instructor in 1899 and was granted his Ph.D. in 1903. In 1910 he was promoted to associate professor. In 1901, while still working on his doctorate, he was appointed Assistant Geologist in the U. S. Geological Survey. In 1909, he was promoted to Geologist, a rank he held until his death.

While on the geology staff at the University of Chicago, Atwood started his field work in the Rocky Mountains, an interest which continued throughout his life. His Rocky Mountain field research lead to the publication in 1932 of *Physiography and Quaternary Geology of the San Juan Mountains, Colorado* in collaboration with K.J. Mather. He also published numerous technical articles in the field of physiography. In 1945, one year before his retirement, he contributed a popular volume on the Rocky Mountains to the American Mountain Series.

In 1913 Harvard University invited Professor Atwood to occupy the chair vacated by the retirement of William Morris Davis. Atwood became interested in writing grade school textbooks while at Harvard; from 1919 on he authored and coauthored a long list of textbooks. Even after officially retiring, Atwood's dynamic personality did not allow him to become inactive. During his last days he was planning a new book for junior high schools entitled *An Introduction to World Understanding — A Social and Economic Geography.* Many believe The Physiographic Provinces of North America was his best university-level textbook.

In 1920 Atwood accepted an invitation from the Board of Trustees of Clark University to start a Graduate School of Geography and to also become president of the University. He brought to the faculty many fine scholars who attracted hundreds of graduate students from more than a score of foreign countries. Atwood concentrated his talents and energy in developing the School of Geography.

Atwood was not only a leader but a loyal worker in many professional organizations. He participated in the National Council of Geography Teachers from its inception and in 1921 became its president. He was also president of the Association of American Geographers and of the National Parks Association.

Atwood's many honors included a Service Award from the University of Chicago and the National Council for Geography Teachers Distinguished Service Award in 1943. The Geographical Society of Chicago presented him with the Helen Culver Gold Medal in 1948 for "distinguished service in the field of geomorphology." Both Clark University and Worcester Polytechnic Institute conferred honorary degrees on him.

Perhaps the estabiishment in 1920 of Clark University's Graduate School of Geography stands as Atwoods' most conspicuous achievement. Its tremendous role in the training of professional geographers in the 1920s was largely due to him. Professor Atwood's life and scientific publications attest to his belief that true geographic training begins in the field. Under his administration the annual fall field camp was an integral part of the graduate program at Clark University.

With all his degrees, decorations and awards, Atwood remained warm and unassuming. His students remember his modest dignity, his sterling character, and his outstanding kindliness (Cressey, 1949:298).

Nels August Bengtson

Nels August Bengtson
May 22, 1879 - 1963

Nels August Bengtson made important contributions to the science of geography and geology, to the teaching of geography, and to liberal education.

He felt a deep commitment to scientific research and to the sharing of information with students and the public through good teaching. As the first dean of the Junior Division at the University of Nebraska, he accepted the challenge of introducing incoming college students to a broad program of liberal education (Hewes, 1964:85).

Bengtson was born in the small village of Winslof, Sweden. His parents migrated from Sweden to a Nebraska homestead in 1881 when he was two years old. He became a naturalized American citizen in 1894. He was ever grateful to both countries and served both his adopted country and northwest Europe well. He served on the U.S. War Trade Board in 1918 and as a Trade Commissioner to Norway. He served in the Bureau of Soils, U.S. Department of Agriculture (1914-18), and wrote eleven reports for the Department of Commerce which were published in 1919. He served as Commissioner in the U.S. Department of Commerce in 1919.

At the age of twenty-one, Bengston entered the Normal School at Peru. He graduated from the two-year program in 1902, married the same year, and taught science for a year at Wood River High School in the Platte River Valley. In 1904 he returned to the Peru Normal School as an assistant in the physical science department. Among other subjects, he taught a beginning course in physical geography.

In October 1905 he enrolled at the University of Nebraska where he received his B.A. degree in 1907 and M.S. in 1908. He served as an instructor (1908-09), assistant professor (1909-12), became associate professor of geology and geography in 1913, and was appointed full professor in 1915. He took a leave of absence during World War I. After his return in September 1920, a non-teaching division of Conservation and Survey was organized under the direction of Dr. George E. Condra, bringing geography back into the Department of Geology.

In the summer of 1917 Bengtson worked as an oil geologist. After returning from Norway he worked as a geologist in Honduras in 1920 and in Ecuador in 1922. Later he served as an oil geologist in Venezuela.

In 1927 Bengtson received his Ph.D. degree from Clark University. In 1929 a separate Department of Geography was established at Nebraska under his chairmanship.

In 1930 Bengtson invited William Van Royan to join him in writing a book on economic geography with an emphasis on resources, designed to be used primarily for college and university students. *Fundamentals of Economic Geography* went through five editions before Bengtson's death.

Coauthor Van Royan wrote:

> To a considerable extent *Fundamentals* grew out of Bengtson's teaching of a large two-semester beginning course mostly for students who were not going to specialize in geography. This course he loved and no other member of the staff was allowed to touch it even though it had to run in several sections. Three-hour weekly laboratories taught by other members of the staff and graduate student assistants supplemented the lectures, as did field trips (Van Royan, 1968:602).

Bengtson also returned during the summers to teach economic geography at Columbia University. Bengtson was not only an unusually good teacher; he also tried to foster good teaching in others, especially graduate assistants. He listened intently to their teaching problems and made suggestions on how to solve them. At one period in his career he was strongly tempted to leave academic work and become a full-time oil geologist. A dean, in trying to persuade him not to leave the University, inquired, "After you are dead, Nels, what do you want on your headstone? Here lies a good rack hound or here lies a great teacher?" Scores of graduate students and thousands of undergraduates are grateful that he chose the latter.

Bengston's interest in his students was sincere, widespread, and devoted. Much of his life was spent teaching, counseling students, and introducing incoming college students to a broad program of liberal education. His interest in and commitment to a student's education was undoubtedly largely responsible for his appointment as dean of the Junior Division of the University of Nebraska in 1940 while also chairman of the Geography Department. He retired in the latter capacity in 1946, as dean in 1948, and as professor in 1949.

After geography became an independent department in 1929 (the first geography course was given at the University of Nebraska in 1903 under a geology number), the graduate geography program was expanded so that in the early 1950s:

> . . . the University of Nebraska ranked fourth in the number of M.S.'s granted to members of the Association of American Geographers and seventh in the number of Ph.D's held by the membership (Hewes, personal correspondence).

Bengtson devoted so much of his time and energy to being an inspiring teacher, a genuinely sympathic and helpful consultant, and efficient administrator that some regret that "Bengtson never was able to set aside time for non-educational profession publication" (Van Royan, 1968:603).

Van Royan, an internationally known geographer who worked more intimately with Bengtson than any other geographer, believes that

"Bengtson could have written an excellent and scholarly work on the geography of the central plains area" (Van Royan, 1968:603).

Bengtson had strong philosophical beliefs concerning his speciality. He believed that for all professional geographers field work was a requisite in their training and important to their further development. He looked at geography as ". . . one of the few *traits d'union* between the more pure physical and the social sciences, and a field which could not and should not be forced into either one or the other pigeon-hole" (Van Royan, 1968:603-605).

The molding of his own career had not been easy, and he had deep sympathy with graduate students who also struggled financially. When I (Barton) first enrolled at Nebraksa it was with some reluctance that I replied to his question, "Where are Mrs. Barton and you living?" I gave him an address on the wrong side of the railroad tracks. His comment was, "With lower rent you'll have more money to buy beans." He was constantly informing graduate students that the goal of an advanced degree was sufficiently worthwhile to justify a diet of beans.

Recognition came to Nels Bengtson in many ways. He was vice-president of the Association of American Geographers in 1923 and in 1942. He served as editor for Prentice-Hall, Inc. He was also a leader in the National Council of Geography Teachers during 1916-35 (Barton, 1964:353). He was a member of the first Executive Committee and of the Board of Directors. He served as national treasurer (1918-19), first vice-president (1927), and president (1929), and was elected to the Executive Committee during 1930-32. Bengtson not only published two major papers in the *Journal of Geography* during this period, but also set an example for his graduate students by giving equal respect to the National Council of Geography Teachers, the American Geographical Society, and the Association of American Geographers. Bengston and his staff supported the Nebraska Council of Geography Teacher. He encouraged this writer (Barton), as a graduate student, to read a paper before this state organization and later to accept its presidency.

Professor Bengtson was president of the Nebraska Academy of Sciences in 1935. He was a member of and held offices in Phi Beta Kappa and Sigma Xi. As an oil geologist, he did research for large international oil companies and was repeatedly employed in field work in Central America and the northern part of South America. Because of his training, field work, employment and publications, Bengtson considered himself both a geographer and a geologist.

Bengtson served as visiting professor at the University of Missouri the year after his retirement, and as acting head of the departments of physics and sociology, University of Nebraska, from November 1952 to November 1953. He also remained as official advisor to graduate students in the Geography Department.

Isaiah Bowman

Isaiah Bowman
December 26, 1878 - January 6, 1949

Isaiah Bowman was born in Waterloo, Ontario, Canada. When he was eight weeks old, his family moved to a log cabin in Michigan, where Isaiah grew up with an early interest in natural history and geography. This led him to the Ferris Institute and to the State Normal School at Ypsilanti. After the formative influence of Mark Jefferson, Bowman went to Harvard University to train under William Morris Davis. Bowman spoke frequently and affectionately of the rigors and values of Davis' training and the influence of Jefferson.

Bowman, who was Director of the American Geographical Society for twenty years and President of Johns Hopkins University for the following thirteen years, died in Baltimore a few days after his seventy-first birthday.

Raye R. Platt writes, "No geographer was called upon by his government so frequently or served it so untiringly or fruitfully (Platt, 1950:174).

Bowman served as Chief Territorial Advisor to the American Commission to Negotiate Peace after World War I and as a member of the Policy Committee of the Department of State. He served as Chairman of the Department of State Territorial Committee and was Special Advisor to the Secretary of State in World War II. He was a member of the U. S. Delegation to the Dumbarton Oaks Conference and Advisor to the U. S. Delegation to the United Nations Conference in 1945, and was Chairman of the Committee on Overseas Territories of the Economic Cooperative Administration.

In 1936 President Bowman was awarded the National Council for Geography Teachers Service Award. Alice Foster, Chairman of the Award Committee who prepared the citation, listed the following items to substantiate Dr. Bowman's leadership in geographic education during a period when the subject was undergoing a rapid transition:

1. Numerous articles published in the decade before World War I indicate his pioneer work in "humanizing geography," a change which the educational world was pleading for at the time;

2. He showed an early interest in geographic education by acting as contributing editor of the Journal of Geography from 1907-1918;

3. He contributed much to geographic education at an adult level through his distinguished work as Director of the American Geographical Society;

4. Through the scholarly works which gave the results of his research and exploration, he has made a valuable addition to the literature available for studying the little known parts of the world;

5. Through the timely publication of *The New World* made possible by his work in connection with the Peace Conference, he facilitated greatly the teaching of geography during the early post-war period;

6. The preparation of the one-to-a-million map of Hispanic America, undertaken by the American Geographical Society under his directorship, represents an invaluable contribution to the material available for geographic education;

7. His service as president of the International Geographical Union from 1931 to 1934 testify to an interest in geographic education extending beyond national boundaries;

8. He rendered valuable services to geographic education as a member of the Krey Commission, particularly through the monumental volume, Geography in Relation to the Social Sciences, which he contributed to the report of the commission" (Foster, 1937:71-72).

The specific services itemized in the award citation were rendered soon after the end of World War I, when Bowman's clear vision and gracious action gave the National Council of Geography Teachers an official publication. In 1918, under Bowman's directorship, the National Council of Geography Teachers made arrangements whereby it assumed ownership of the *Journal of Geography*. Previously the *Journal* had been under separate management and control.

During 1919 the *Journal* was published for the first time as the official publication of the National Council of Geography Teachers, an organization some five years old. This change in ownership came about at the suggestion of Bowman, whose name is listed on the title page of the 1919 volume as editor. From January 1919 to May 1920, the *Journal* was published in New York City by the American Geographical Society, with Bowman as editor. In May 1920, the National Council assumed control of the *Journal*, with George J. Miller as editor and J. Nystrom as publisher.

A speaking engagement prevented President Bowman from attending the 1936 annual dinner of the National Council of Geography Teachers, where the NCGT Distinguished Service Award was to be made. However, he sent a letter in which the following appeared:

In accepting your award and thanking you for it I beg to express the hope that the National Council of Geography Teachers will reach higher and higher standards of teaching and that it will never consent to look upon teaching as something apart from research but, rather, that teaching and research shall be regarded as the right and left hands of the educational process (Foster, 1937:72).

A colleague characterized him thus:

> Isaiah Bowman was a man of great energy, great decisiveness, and penetrating intellect . . . His boundless energy he carefully channeled so as to be able to do more things than several ordinary men could hope to accomplish" (Carter, 1940:335-350).

Bowman's knowledge and wisdom made him a popular and valued consultant to presidents, statesmen, faculty members and students. He was a great administrator, due in part to his ability to swiftly penetrate a problem, reach a correct decision, and set in motion machinery to accomplish his aim. Although he died nearly a quarter of a century ago, his published works (Carter, 1950:338-350) remain to inspire us as we strive to improve the field of geography.

R.D. Calkins
April 20, 1873 - September 27, 1955

R.D. Calkins was born in the village of Parma in southern Michigan, ten miles west of Jackson. His early life was uneventful and resembled that of thousands of aspiring youngsters of his time. He attended a rural elementary school and a small town high school. He then taught in the public schools of his home community to earn money for his college education. Calkins acquired his interest in geography through reading N.S. Shaler's "Nature and Man in America" in *Scribners Magazine*" (Leighly, 1956:43).

Calkins's college career started at the State Normal School in Ypsilanti. He then transferred to the University of Chicago, from which he received a B.S. degree in 1903 and an M.S. in 1907. In one interval between periods of study at Chicago, he was an instructor at Ypsilanti for two years (1898-1900), teaching under the supervision of his former professor, Mark Jefferson. Calkins went from Ypsilanti to the State Normal School (now Central Michigan University) at Mount Pleasant, where he served until his retirement in 1943.

His undivided devotion to the tasks at hand and a modest nature prevented Calkins from becoming as widely known as many less able colleagues (Leighly, 1956:43). He was at his best in the classroom, which was admirably equipped in the departmental quarters built in 1909, according to his plans.

Leighly reports that:

> Before a class his shyness and self consciousness dropped away as he bent his great physical and mental powers to the task of leading students through chains of reasoning to the answers of questions that when first posed often appeared to the students beyond his ability to answer (Leighly, 1956:43).

Calkins was fortunate to have studied under two great geography teachers — Mark Jefferson at Ypsilanti and Rollin D. Salisbury at Chicago. Of the latter Calkins once wrote:

> He did not lecture. He taught. Under his skillful questioning geology developed in our minds as we thought our way through the science.

Calkins might have been describing himself since he applied the same teaching methods in his classroom.

Professor Leighly reports that for Calkins:

> The objective fact was as prominent in his teaching as was the logical inference. Under his guidance students learned to know the physical features of the country about Mount

Pleasant. He had in his department and used with consummate skill an unsurpassed collection of teaching materials from the "dingus" on his lecture table with which he demonstrated the path of the sun through the sky in different latitudes to cabinets of minerals, rocks, and commercial products. Those of his students who went on to advanced work in the universities soon learned that the instruction Calkins had lead them through was more rigorous than much they encountered later (Leighly, 1956:43).

On writing for publication, Professor Calkins said:

Why should I write about the little things I have observed. Let someone else do that. My job is to improve the teaching in the public schools of Michigan (Leighly, 1956:43).

Calkins accomplished his goal. What our geographic legacy would have been had he been a prolific writer can only be surmised. Sauer is reported to have said of Calkins, "He never stopped learning" (Stanislawski, 1975:554).

William Morris Davis

William Morris Davis
February 12, 1850 - February 5, 1934

William Morris Davis was the Dean of American Geographers. He was a leader for fifty years in a new profession called geography, which was established in the U.S. during the last part of the nineteenth and the early part of the twentieth centuries. According to Richard Elwood Dodge:

> During his long life of productive investigation and publication he exerted a greater influence on scientific geography and on educational geography from the elementary school to the university than has any other American scholar (Dodge, 1934:148-150).

When he received the Penrose Medal for his contributions to geology — the greatest recognition that can be given to a geologist — Davis said that he followed two maxims, "go and see" and "see and think" as he traveled to every continent except Antartica. Davis was the first to organize the principles of meteorology, and for many years his textbook was the standard reference volume on meteorology in the U.S.

Davis was born into a prominent and influential Quaker family in Philadelphia. His grandmother, Lucretia Mott, was a formidable bluestocking who campaigned for women's rights, founded the American Anti-Slavery Society, read only mind-improvement books, and abhorred alcohol. Davis was tutored at home during his early years, then at sixteen entered Harvard University, where he received two science degrees.

In 1870 Davis began a three-year term as assistant in the Meteorological Service of the National Observatory at Cardoba, Argentina, located on the high plains sloping eastward from the Andes (Martin, 1950:175-180). This experience introduced him to the world which would become his classroom, laboratory, and field of investigation for life.

Davis traveled around the world in 1876-77; to Turkestan with Raphael Pumpelly, an influential Harvard professor, in 1903; to South Africa in 1905; to Australia by way of Hawaii and numerous small Pacific Islands in 1914; and to the coral reef islands of the Lesser Antilles in 1923.

Interspersed with these trips, Davis served as visiting professor at the University of Berlin (1908-09) and at the Sarbonne of the University of Paris (1911-12). During these teaching assignments, he studied European geomorphology and lectured on it and on the physiography of the United States. Along with his classroom teaching, Davis visited areas of Germany and France where geomorphic features intrigued him. He also conducted a nine-week geographical field trip from Ireland to Italy in

1908, and later led an excursion in the Alps and in northern Italy. In later years he revisited some of these areas with his American students.

In 1912, at the age of sixty-two, Professor Davis planned, financed, and conducted the famous transcontinental excursion of the U. S. This tour brought to America forty very distinguished and several promising European geographers. During two months these guests were taken through representative areas of the U. S. (with the exception of New England), traveling by automobile, train, stage coach, steamboat, and on foot, with stops wherever geomorphology beckoned under Davis's guidance.

Professor Davis possessed an unusual linguistic ability for an American geographer. His university lectures in Germany and France were conducted in English, but he spoke German and French quite well. He undoubtedly acquired fluency in Spanish during his three years in Argentina, and it was reported that he had a reading and speaking knowledge of Italian. His books were translated into German, Italian, and Japanese. Davis read geomorphological, geological, and geographical papers and books in European languages rather than depend on translations or the English summaries.

After serving in the Meteorological Observatory at Cordoba, Davis returned to Harvard and was assistant to and instructor under Nathaniel Southgate Shaler in the Department of Geology until 1885. He was assistant professor and then professor of physical geography until 1899, when he became the Sturgis Hooper Professor of Geography (Bryan, 1935:23-31). After a thirty-five year tenure at Harvard, Davis retired, but he continued teaching as a visiting lecturer at Columbia, Stanford, Oregon, Arizona, and the California Institute of Technology.

During his lifetime many honors were bestowed on Davis, including honorary doctoral degrees from several universities in Europe, Australia, and South Africa. He also received numerous medals from geological, geographical, and other learned societies.

Davis made numerous and varied contributions to geography and science. N. S. Shaler and Davis collaborated on a book, *Illustrations of the Earth's Surface,* which was published in 1881. Davis developed and used his own method of showing ideas in drawings that took the form of perspective landscapes and block diagrams in publications and teachings. Using a unique combination of graphic skill and analytical analysis, Davis made the use of block diagrams universal.

Between 1884 and 1901 Davis published a number of articles on meteorology. In 1894 he published his famous and much used textbook, *Elementary Meteorology*. He analyzed and explained the research of the specialists with such clarity that the basic principles were easily understood by young students. His great contribution in geomorphology was the formulation and elaboration of the doctrine of the Cycle of Erosion (Baulig, 1950:188-195).

Roderick Peattie, who was a faculty member at Ohio State University, writes about Davis, "He was indeed the father of American geography. He defined the science and was merciless in making us stick to that definition during his life time" (Martin, 1950:178).

The Journal of Geography, the official journal of the National Council for Geographic Education, indirectly owes its existence to Davis. Richard Elwood Dodge, a student of Davis at Harvard, established the *Journal of School Geography* because of Davis's initiative and insistence (Dodge, 1934:148-150). *The Journal of Geography* became a successor to the *Journal of School Geography* in 1902.

Isaiah Bowman characterized Davis, his professor at Harvard, as "a born teacher of scholars, forever sowing the seed" (Bowman, 1934:177-181). Prominent among Davis's students were Bowman, Richard E. Dodge, Robert C. Ward, Ralph S. Tarr, Albert Perry Brigham, Walter W. Tower, Walter C. Mendenhall, Philip Sidney Smith, and Walter Goldthwait.

In 1932 William Morris Davis received the first Distinguished Service Award given by the National Council of Geography Teachers in recognition of outstanding contributions to geographic education.

Richard Elwood Dodge

Richard Elwood Dodge
March 30, 1868 - April 2, 1952

Richard Elwood Dodge earned national recognition as a teacher, author, editor, administrator, and creative organizer in the field of educational geography. He was the founder and first editor of the *Journal of Geography*, the official journal of the National Council for Geographic Education. During his eighty-four years he made numerous and varied contributions to geography and education.

Richard E. Dodge was the first professor of geography at the Teachers College, Columbia University. He pioneered in the cause of human geography and was at the same time a reputable physical geographer. The lion's share of his contribution to educational and human geography was during his twenty-one fruitful years on the faculty at the Teachers College.

Dodge was born on a farm in Wenham, Massachusetts. His early, rural experiences stimulated his interest in geography and proved of great value to him in his later studies.

On July 4 1876 Dodge attended the famous Centennial Procession at Boston which commemorated the first 100 years of American independence. It was his first contact with events and affairs outside his own small world. Attending Harvard, he received the A.B. degree in 1890 and the A.M. degree in 1894, after which he was appointed an instructor in geology there. Dodge was fortunate to attend Harvard when he did because he could enroll in the classes of such great scholars as Nathaniel S. Shaler, Albert P. Brigham, R.S. Tarr, Robert DeCourcey Ward, A.H. Brooks, and William Morris Davis. As he worked with these noted scholars, Dodge began to find an intellectual outlet in the geographic discipline.

After extensive summer field work, including participation in the famous Transcontinental Excursion of 1891, Dodge decided he would be a geographer rather than a geologist. In making this decision, he owed much to the early guidance of Davis and Shaler.

At Teachers College, Columbia University, whose faculty he joined in 1895, Dodge was primarily concerned with training geography teachers. His experiences led him to establish the *Journal of School Geography* in 1897, which he edited for thirteen years (Griffin, 1952:253-258). The *Journal of Geography* soon became the most influential publication in the field of educational geography.

His editorial abilities were recognized also by the Association of American Geographers. He was selected as the first editor of the *Annals*, serving from 1911 to 1923, except for one year (1915) when Harlan H.

Vernor Clifford Finch

Barrows served as editor while Dodge served as president of the Association. None of his successors, except Derwent S. Whittlesey, served so long as the *Annals* editor. Dodge's long term as the *Annals* editor was in recognition of his expertise in judging the acceptability and research value of the works of others.

With Isaiah Bowman as editorial associate and I.C. Le Compte as translator, Dodge published the work of the eminent French geographer, Jean Brunhes, *La Geographie Humaine,* in English in 1920.

Richard Dodge wrote numerous geographical papers and a number of textbooks for both elementary and college students. In 1903 they appeared as the Dodge Geographies, followed in 1927 by the Dodge-Lackey Geographies. Ten years later, in 1937, he and his son published a college text, *Foundations of Geography,* and with N.H. Carter he published *Economic Geography* in 1939.

After serving twenty-one years on the Teachers College faculty at Columbia, Dodge brought geography to the Connecticut State College at Storrs, where he was dean from 1920-1930.

In 1946 the National Council of Geography Teachers recognized Richard Elwood Dodge's successful and long career in geography and geographical education by presenting him with its Distinguished Service Award. The citation stressed his interests in the wider field of science. He was a Fellow of the New York Academy of Science, the Geological Society of America, and the American Association for the Advancement of Science. He was also a member of Sigma Xi and active in Rotary International.

Vernor Clifford Finch
October 18, 1883 - October 23, 1959

Vernor C. Finch was born on a farm near Tecumseh, Michigan. He graduated from Kalamazoo College in June 1908, and received a bachelors degree from the University of Chicago during the same summer. After a year of high school teaching at Mt. Pleasant, Iowa, he returned to the University of Chicago for two years of graduate work in geography.

In 1911 he was appointed assistant in geography at the University of Wisconsin. He became an instructor in 1912, assistant professor in 1916 (following completion of his Ph.D. during that year), associate professor in 1918, and professor in 1927. In 1928, when geography was separated from geology, he was appointed chairman of the new department and served in that position for seventeen years until he asked to be relieved in 1945. In spite of ill health, Professor Finch continued to carry a full share of departmental work, but in 1951 failing health forced him to retire. In 1954 he was named professor emeritus.

Highlights in Finch's career, as evaluated by Charles C. Colby (1960:239), a lifelong fellow geographer, include:

> (1) his cooperation with O. E. Baker, under the auspices of the U.S. Department of Agriculture, in the production of the famous *Geography of World Agriculture.* This volume was used for decades as a basic work in the study of agricultural geography; in addition, it was long recognized as a model in geographic cartography in its use of the dot map technique to portray distributions on a macroscopic scale (Hartshorne, 1961: 339-342);
>
> (2) his two years of service during World War I as special expert in the Commodity Section of the Division Planning and Statistics of the United States Shipping Board.

Two of Professor Finch's research publications have been recognized as landmarks in the development of American geography (Hartshorne, 1961:340). The first was the production of the *Geography of the World's Agriculture,* already referred to above; the second was the "Montfort Study." His demonstration of the "fractional code" method of representation of landscapes brought to a conclusion a series of discussions and experiments by a number of prominent American geographers. The definitive demonstration of both the advantages and disadvantages of the technique provided the basis on which later students could rate the feasibility of applying the technique to research practical problems of regional planning. The Montfort Study was financed by the Wisconsin Historical Society and published by the Geographical Society of Chicago.

As a modest and unassuming man, Finch was rather embarrassed by the honors he received. In 1930 the American Geographical Society elected him to an honorary corresponding membership. In 1933 he was granted an honorary Doctor of Science degree by his alma mater, Kalamazoo College. In 1938 he was named president of the Association of American Geographers. In 1948, he received the Helen Culver Gold Medal of the Geographical Society of Chicago "for distinguished contributions to Economic Geography." In 1950 he was awarded the Distinguished Service Award, the highest honor granted by the National Council of Geography Teachers. Following his retirement, his office was converted into the Finch Seminar Room. In 1961, two years after his death, the University of Wisconsin named one of the new professorships to be established for research the Vernor C. Finch Research Professorship.

To numerous graduate students, including Tom Barton, who served as his assistant and student in his "Anglo America" and "Geographic Regions of the South" seminar courses, Finch was the epitome of scholar, gentleman, and friend.

Alice Foster

Alice Foster
December 4, 1872 - July 7, 1962

Alice Foster was born in Ashton, Iowa. She first taught for several years in Mexico, later published several articles about Mexico, and wrote her masters thesis on Mexico.

Upon returning from Mexico, Foster began her college career. After receiving an A.B. from the University of Iowa in 1918, she joined the faculty of Iowa State Teachers College, where she taught from 1919 to 1920. She then went to the University of Chicago, where she received an S.M. in geography in 1921. She taught at Mount Holyoke College from 1921 to 1926. Returning to Chicago, she taught in the Laboratory School of the University of Chicago during 1928-32. She did her graduate work at the University of Chicago, completing in 1936, at the age of sixty-three, her doctoral dissertation on Spain, entitled "The Geographic Structure of the Vega de Valencia."

A crucial point in Foster's career as a geographer was her association with Charles C. Colby. Several of her articles on Mexico were included in his *Source Book for the Economic Geography of North America* (1921). They later combined their talents to write an *Economic Geography for Secondary Schools* (1931) and a much revised and enlarged *Economic Geography: Industries and Resources of the Commercial World* (1940). For each of these, she produced a workbook of directed studies or investigation (Harris, 1963:77).

Foster provided astute leadership in the field of geographic education. In 1941, while serving as president of the National Council of Geography Teachers, she realized that the organization's activities since its inception over four decades earlier should be recorded. To meet this need, in 1941 she prepared a monograph entitled *The National Council of Geography Teachers as a Working Organization* from records of the NCGT and the recollections of George J. Miller. Foster was the director and driving force of the project, and Thomas Frank Barton, as the recently elected secretary of the organization, aided in compiling the information. The monograph provided an accurate, historical document which aided in holding the organization together during World War II, when annual meetings were not held, and formed a basis on which to revitalize the organization when the war ended.

Among Foster's other educational geographic contributions were chapters in the *32nd Yearbook of the National Society for the Study of Education,* in the *1948 Yearbook of the National Council for the Social Studies,* and five articles in the *Journal of Geography.* She also contributed papers to the three leading geographical periodicals: the

John Paul Goode

Annals of the Association of American Geographers, the *Geographical Review,* and *Economic Geography.*

Foster served as summer lecturer at California State Teachers College, San Diego (1924), the University of Chicago (1936, 1937), Western Reserve University (1938), and Kent State University (1939). She specialized in geography in education, the geography of Spain, tropical America, and economic geography.

During World War II Foster returned to active teaching at the University of Chicago as instructor in the Premeteorological Program for Air Force Cadets and later served as geographic examiner for the Armed Forces Institute.

One of her colleagues gave this impression of Foster:

> Alice Foster was loyally and warmly devoted to family, friends, professional colleagues, and the field of geography. She had a keen interest in young people and thereby seemed always to be young herself. Literally hundreds of students remember her with the kindliest of affection for her sympathy, encouragement, and helpful advice over the years. Long before formally becoming Alumni Secretary of the Department of Geography of the University of Chicago in 1950, she was formally advisor and friend to graduate students and former students (Harris, 1963:77).

In 1947 the National Council of Geography Teachers presented Foster with its Distinguished Service Award.

John Paul Goode
November 21, 1862 - August 5, 1932

John Paul Goode was born on a farm near Stewartville in southeast Minnesota. He was the product of an austere pioneer life in a land of long, cold winters. Late in life he was fond of relating his struggle to get an education. He worked his way through preparatory school and the University of Minnesota. Gifted with a beautiful voice, he earned part of his living expenses by singing in choirs and quartets. During the last few months of his life, although constantly aware of a weakened heart, he formed a small orchestral group made up of boys and girls from the neighborhood who loved him and considered him a friend who understood their ambitions and dreams.

In 1889, after receiving a B.S. from the University of Minnesota, he was called to teach natural science at Minnesota State Normal at Moorhead, where he remained until 1898. He interrupted these years of teaching to do graduate work in 1894 at Harvard and to serve in 1896-97 as a fellow at the University of Chicago. John Paul Goode developed so rapidly during these early years of teaching that he was well prepared to teach geography classes at the University of Chicago during the summers of 1892-1900. In 1899 he accepted a position as professor of physical science and geography at Eastern Illinois State Normal.

Graduate study was so enticing that in 1901 Goode entered the University of Pennsylvania, where he earned a Ph.D in economics. He remained at the University of Pennsylvania until 1903 as an instructor in economic geography and director of the Geographical Society of Philadelphia.

Partly because he had already identified with the University of Chicago, in 1903 Goode was offered a permanent position in its newly established Geography Department. The Chicago department grew rapidly and Professor Goode worked diligently to develop geographers into qualified teachers.

Goode's influence extended far beyond the classroom. He served as coeditor of the *Journal of Geography* from 1901 to 1904, helped organize the Geographic Society of Chicago, and was its president (1904-06) and remained a director until 1929. He acted as general secretary of the American Association for the Advancement of Science during 1907-08.

Since he was an expert in economic geography, in 1908 the Chicago Harbor Commission sent him to Europe to study the leading ports. His complete and extensive report was published under the title, "The Development of Commercial Ports." The following year he was appointed by President Taft to assist a distinguished group of Japanese financiers, the Honorary Commercial Commissioners, in a tour of America, visiting fifty-five leading cities. His interest in the Far East led to an invitation by the Philippine government in 1911 to lecture to the Educational Assembly at the summer capital at Baguio. Goode then traveled and studied in the Philippines, China, and Japan for the remainder of the year.

Goode's students were employed in many fields. They remembered the inspiration they received from his classroom lectures. Invitations poured in to Goode, requesting him to speak before teachers associations, community assemblies, business groups, and other organizations. With characteristic thoroughness, energy, and humor he developed the art of public lecturing to the highest degree. During 1926 he filled seventy engagements. Each lecture, illustrated by slides of graphs and maps, was filled with striking vividness and charm.

Goode was also active in the Association of American Geographers. He presented two addresses at the annual program of the AAG in 1907. At the Baltimore meetings in 1908 he launched what became a series of papers on cartography with one entitled "The Requisities of a School Wall Map." He led the fight against the Mercator Projection for school maps; he deplored the fact that the Mercator Projection was used so extensively. He started his own great work on producing an equal-area projection that could be used for world maps.

He appeared eleven times on the AAG programs after the Baltimore meeting and at each appearance, with one exception, he noted the advances made in cartography. Five of these appearances were landmarks in his career as a progressive cartographer. In 1914, at Chicago, his address was entitled, "A New Series of Wall Maps for Schools." In 1916, in New York City, he spoke on "New Idea for a World Map: A Substitute for Mercator's Projection (Mollweide's Homolographic with Interruption)." In 1923, in Cincinnati, he spoke on "The Homolosine Projection." In 1928, in New York, he lectured on "The Polar Equal Area (A New Projection for the World Map)" (Haas and Ward, 1933:241-246).

Of great importance were Goode's contributions in 1916 and in 1923. The interrupted homolographic represented a distinct advancement in cartography and the homolosine project was hailed as the solution to the problem with which he had wrestled so many years. These two projections ushered in a new epoch in mapping and were the inspiration which led to many other efforts. The two projections, or modifications of them, were adopted in all important cartographic institutions and in many government bureaus.

During the decades that Goode was teaching, lecturing, and working on wall maps, his chief ambition was to produce a school atlas that would stand as his greatest contribution to cartography and to American geography (Haas and Ward, 1933:245). His preliminary *Atlas,* published in 1923, was a great source of satisfaction, and he felt the climax of his life's work was reached when the enlarged and revised edition of the *School Atlas* was completed shortly before he died.

Official recognitions of Professor Goode's significant contributions to geography were varied and numerous, but his chief pleasure in life was the loyality of his students and friends. The closing years of his life were fraught with discouragements and a deep sorrow from which he never recovered. His wife passed away on January 13 1928. Goode had collapsed with a heart attack on December 11 and was unable to help or see Mrs. Goode during the last four weeks of her life. This was a severe blow, since she had been his helpmate, companion in his triumphs, and inspiration and guide during many critical periods.

Gilbert H. Grosvenor

However, with characteristic determination and a driving will, Goode recovered from the December heart attack and was back at work the following spring with instructions from his doctor to "go slow — do only one man's work." During his last, brief illness during the summer of 1932 he received many messages of congratulation on the publication of his *School Atlas.*

Gilbert Hovey Grosvenor
October 28, 1875 - February 4, 1966

An account of the years of Gilbert Grosvenor is the history of the National Geographic Society, the organization with which he was associated throughout his adult life.

Gilbert Grosvenor was born of American parents in Istanbul (then called Constantinople), Turkey. At the time his father, Edwin Augustus Grosvenor, was professor of history at Roberts College. In 1891, when the family returned to the United States, the father joined the staff of Amherst College, where in due course Gilbert enrolled as a student. In 1897 Gilbert graduated magna cum laude with a bachelor of arts degree. In 1901 he received his masters degree and also married Elsie May Bell, the daughter of Alexander Graham Bell, the inventor of the telephone.

In 1888 the National Geographic Society was founded for "the increase and diffusion of geographic knowledge." The Society, although small in membership, published a journal which had financial difficulties in the early years. Because of Grosvenor's interest in the organization, Bell asked him to become editor of the Society's journal.

On April 1 1899, when Grosvenor became editor, his headquarters consisted of half a rented room on the fifth floor of a Washington, D.C. office building (Wetmore, 1966:598-599). At first Dr. Bell contributed one hundred dollars monthly toward expenses, a sum which paid the editor's salary. Grosvenor made efforts to increase membership in the Society so that it would be self supporting.

Grosvenor studied the standard works on geography and eventually developed an editorial policy based on two principles: (1) articles should be based on accurate eyewitness accounts and presented in a simple language — "writing that sought to make pictures in the reader's mind"; (2) supplement word pictures with ample photographs and maps. Since suitable illustrations were difficult to locate, Grosvenor began to take photographs pertinent to his articles.

At the end of 1904, Grosvenor, in his layout of the *National Geographic* magazine, found he lacked eleven pages of copy to fill an issue. He received in the mail some excellent photographs of Lhasa, a little known city in Tibet known to the Western world as a forbidden land. Grosvenor, with some trepidation, ran the extraordinary pictures, which were received with enthusiasm. A few days later his board of managers (now called trustees) unanimously elected him to the board. Photographs became a regular part of each issue of the magazine; six years later color illustrations were added. Maps became another standard addition. Some of the articles were expanded and appeared in book form. These editorial practices steadily attracted additional membership in the Society.

By 1907 the organization could afford to offer modest grants toward geographic research. The first grant was a contribution of $1,000 toward the expenses of Robert E. Peary's expedition to the North Pole. One of the recent notable scientific achievements made with the Society's economic support was the creation of the *National Geographic Palomar Sky Atlas,* which proved to be a very valuable research tool for the world's astronomers.

In 1963 the Association of American Geographers honored Dr. Grosvenor with its Outstanding Achievement Award. The award reads:

> For more than half a century of successful leadership in the broadening of public appreciation for geography through tremendous growth and circulation of the *National Geographic Magazine,* with its superb illustrations, the production of consistently superior maps, and the support of numerous scientific expeditions and research projects (Ackerman, 1967:197).

Grosvenor, as editor of the *National Geographic,* was a geography teacher to millions of Americans, and millions of pre-high school students' interest in geography was stimulated by reading *School Bulletin,* which was published weekly for school use. Grosvenor was honored with the National Council of Geography Teachers Distinguished Service Award in 1944.

Ellsworth Huntington

Ellsworth Huntington
September 16, 1876 - October 17, 1947

Ellsworth Huntington, Professor Emeritus of Geography at Yale University, died in New Haven at the age of seventy-one. He was one of a few American geographers known throughout the world, becoming famous through his writings, the intellectual debate stimulated by some of his publications, and his broad circle of friends in many varied fields of knowledge. His main interest was the study of the problems of human existence and man's relation to the world (Valkenburg, 1948:153-155).

Huntington was born in Galesburg, Illinois. He received an A.B. degree from Beloit College (1897), an A.M. from Harvard University (1902), and a Ph.D. from Yale (1909). He was an assistant to the president of Euphrates College in Turkey, 1903-04. He served at the Carnegie Institute during 1912-14 and was with R.L. Barrett's expedition to Chinese-Turkestan in 1915-17. After a period of teaching at Yale (1907-13), he became a research associate there in 1917 and, except for being a visiting professor in summer schools at Chicago and Clark universities, he concentrated on research and writing.

Professor Huntington made prolonged and varied field trips in the United States and in foreign countries. After two years at Harvard he carried on field studies in the western United States and Mexico, and later made extensive field studies in Europe, Africa, Australia, and South America.

A colleague and coauthor has written:

> No other geographer has made such prolonged and varied foreign field studies and none has shared his observations so attractively. (Visher, 1947:632-633).

Huntington's early geological work was so outstanding that he was listed as a distinguished geologist in *American Men of Science*. He specialized in the geography of Asia, and early in his career he received the Gill Memorial Medal of the Royal Geographical Society for his work in Asia. He received medals from leading British, French, and American geographical societies for geographic explorations in little known areas in Asia. He received the National Council of Geography Teachers Distinguished Service Award in 1942. He served as president of the Association of American Geographers, the Ecological Society of America, and the American Eugenics Society, and was an associate editor of the

Geographical Review, Economic Geography, Ecology, and *Social Philosophy.*

Perhaps no other American geographer's writings have been read by so many specialists in other fields and by so many college students. Numerous meterologists, climatologists, sociologists, historians, economists, anthropologists, and physicians have frequently cited one or more of Huntington's numerous publications and acknowledged his stimulating influence in their fields.

In evaluating Professor Huntington's success, Van Valkenburg of Clark University stressed four factors:

> First of all, he had a keen mind and the ability to think problems through to a logical solution. Second, he was able to devote himself entirely to research. Third, he could write clearly and simply, sometimes so simply that many were prone to underestimate the amount of painstaking research on which his conclusions were based. As one critic put it, all his books were informed with vitality and humanity. Finally, he knew the world intimately through extensive travel; his photographic mind could recall the minutest details of his trip (Valkenburg, 1948:154).

It is impossible in this small space to mention all his publications and activities. Many of his articles appeared in an array of magazines, and he also contributed chapters in twenty books. He authored or coauthored twenty-six volumes. *Mainsprings of Civilization* (1945) is considered by some scholars to be a monument not only to his exceptional erudition, energy and persistence, but to American science. *Time* magazine, in an obituary, characterized his *Mainsprings of Civilization* as:

> . . . surpassed only by Toynbee's classic in breadth of scholarship, wide interest, and literary attractiveness. (Visher, 1947:633).

In addition to his scholarly works, Huntington was a captain in the Division of Military Intelligence during World War I. In World War II he was called upon to aid the office of the Quartermaster General.

During the spring of 1946 Clark University awarded Huntington a doctor of science degree honoris causa. The citation ended with these words: "The man whose thoughts of today are the geography of tomorrow."

J. B. Jackson

John Brinckerhoff Jackson
September 25, 1909 -

John Brinckerhoff Jackson was born in France. His family was highly educated, well traveled, and had significant social contacts both in Europe and in America. He received an A.B. degree from Harvard University in 1932. During World War II, he was an American intelligence officer. Later he worked on a sheep ranch in the Great Plains of eastern New Mexico. He was attracted to the West and decided to establish his permanent residence near the southern end of the Rocky Mountains in the Hispanic community at La Cienaga.

His home and large personal library at La Cienaga, about a dozen miles south of Santa Fe, is a huge, thick-walled adobe structure set in a secluded spring-watered valley in a grove of immense cottonwood trees. Jackson, one of the few Anglos in this Hispanic settlement, is an active, benevolent, and highly respected member of the community while he maintains an intellectual leadership role on the humanized environment of the national and international scene.

J. B. Jackson founded the internationally acclaimed *Landscape* magazine in 1952 and served as its editor for seventeen years. His brilliant contributions and editorials attracted distinguished scholars and leaders in public policy as writers for *Landscape*. The magazine became renowned among architects, geographers, planners, landscape architects, and laymen who were sensitive to the humanized landscape. The perspective was wide-ranging geographically, historically, politically, and socially.

In addition, Jackson authored such books as *American Space* (Norton & Co., New York, 1972), an analysis of the rural and urban landscape during the crucial period following the Civil War, and *Landscapes* (Univeristy of Massachusetts Press, Amherst, 1970), which is an anthology of his various writings and lectures.

For years, Jackson served as a faculty member at both Harvard University and the University of California, Berkeley. In addition, he has held numerous visiting professorships in geography, such as at the University of Minnesota, and in landscape architecture at the University of Massachusetts. In honor of Professor Jackson, the Association of American Geographers created the highly coveted J.B. Jackson Prize, an annual award for the most outstanding literary contribution to the American scene.

Professor Jackson, long after a normal retirement age, remains one of America's most influential and revered national and international lecturers on landscape as an expression of social values and cultural pat-

Mark Jefferson

terns in the way of life. As a mentor to fledging scholars and founder of *Landscape*, he helped pioneer landscape studies. He has influenced scores of American scholars and authors.

Mark Jefferson
March 1, 1863 - August 8, 1949

Mark Jefferson, who for almost forty years was in charge of geography at the State Normal College, Ypsilanti, Michigan, made a national and international contribution to geography through his publications, teachings, and field studies.

His publications consist of seven books, brochures, at least fifty-five major articles in geology and geography magazines, dozens of reviews, and thirty-one "notes" appearing in books from 1917 to 1938. Each of these "notes," which were less than a page long but represented a huge amount of work, were largely about South America, Europe, world population, and climate (Visher, 1949:307:312).

Some of Jefferson's books and articles were major contributions to the field of geography. Yet writing was a chore for him, according to Isaiah Bowman, who was a protege of Jefferson. Professor Bowman adds, "One of his most charming traits was his desire to tell the world when he had found a new idea or a new geographer whose tracks he had discovered in a journal or book (Bowman, 1950:134-137).

Even with the limited facilites at State Normal College, Jefferson trained his students to become professional geographers and to teach geography with a firm understanding of its basic facts and principles. He kept his eyes open for students with unusual abilities. After a year or two, Jefferson passed along his more talented students to graduate universities. Regardless of whoever their new professors might be, these students never forgot Jefferson, the cultured schoolmaster who had followed their careers with interest and support.

Although Professor Jefferson was repeatedly invited to teach summer sessions in graduate geography departments, he never sought a university chair. After teaching at Harvard University, he taught during the summer sessions at Michigan in 1903, at Yale in 1907, at Chicago in 1917, at California in 1920, and at Columbia in 1934.

Jefferson spent approximately every other summer in field work, either in the United States or in a foreign country. His observations of physical landscapes and man's use of the earth began even before he studied geography academically and continued after he received his degrees and far beyond his retirement.

Jefferson carried on field studies in Europe in 1904, 1911 and 1935, and in South America in 1918 and 1926. In 1911 he accompanied William Morris Davis to Europe, and in 1912 he was marshall for the American Geographical Society's International Transcontinental Tour of the United States.

From November 1918 to June 1919 Jefferson served as Chief Cartographer to the American Peace Commission at the Peace Conference of Paris, directing the making of about five hundred original maps at the Conference. A collection of his Peace Conference maps is included in the American Geographical Society's map collection. According to Isaiah Bowman, who was also a member of the American Peace Commission, because of Jefferson's maps:

> no other delegation or alternative solutions of territorial problems came before the various boundary commissions. Although he had no part in substantive territorial decisions, he was quick to see how they could be graphically represented (Bowman, 1950:136).

Some of Jefferson's South American publications were partly the result of field studies conducted during a 1918 expedition to Argentine, Brazil and Chile, which was financed by the American Geographical Society.

Jefferson's academic training in geography greatly influenced his lifelong career. Until he was twenty years old, he lived in a Boston suburb. Entering Boston University at seventeen, he took classical courses there and specialized in languages, mathematics, and astronomy. During his senior year he was invited to become an assistant astronomer at the National Astronomical Observatory, Cardoba, Argentina, where he spent nearly six years. He became submanager of a sugar estate at Tucuman, Argentina. Between sugar harvests, he returned to Boston, where he graduated with honors in 1889.

After his time in South America, Jefferson returned to the United States and spent about a decade in the Massachusetts public school system as teacher, principal, and superintendent.

In 1901, Jefferson was appointed to the chair of the Geography Department at Michigan State Normal College, Ypsilanti, where he remained until his retirement in 1939. One of his students, Bowman, who became president of Johns Hopkins University, writes:

> To many of the students he was the first broadly cultured man they had met; to some of his faculty members he was a delight; to the looseminded, an annoyance! His keen scythe could mow down a wavering field of words at one sweep. He had a dryness and wit that sometimes offended the dull and that were a constant source of amusement to his friends" (Bowman, 1950:134).

Jefferson's achievements led to high honors. At the annual meeting of the Association of American Geographers, held in Ypsilanti in 1931, he received the Cullum Geographical Medal of the American Geographical Society; in 1932, the Culver Gold Medal. In 1939 he received the National Council of Geography Teachers Distinguished Service Award. In 1902 he became an associate editor of the *Journal of Geography* and in 1909 he became contributing editor of the *Bulletin of the American Geographical Society.*

Clarence F. Jones

Clarence F. Jones
August 30, 1893 - October 12, 1991

Clarence F. Jones was born in Indian Valley, Virginia. In an interview in March 1985 with Karan and Mather, he reported that his father's sister, Virginia, had married a moonshiner down the valley from the Jones's homestead. "At night we would hear him capping the barrels of whiskey. Dad had a good team of mules and a good wagon, and he hauled barrels of whiskey to Christianburg for his brother-in-law. Dad was always careful, but one time he went with two barrels and only one had a revenue stamp on it. After he got home, he got word that the revenue officers were coming. So he beat it for Illinois. He came to Plymouth (sixteen miles southeast of Macomb) and got a job working for a dollar a day. In addition, he got fresh milk and a pig or two for meat. After a year or so, Dad sent for my mother and my brother and me. We were in Illinois for two or three years. My father didn't like it in the Midwest, so we went back to Virginia."

Later the family returned to Illinois, where Clarence attended three different rural schools in McDonough County. Clarence stayed out of school in the fall until the corn was harvested, and stayed out again in the spring to help with the farm work. Because of these absences, he did not pass a single grade of school until he was fifteen years old and in the eighth grade, when he took the final examination and received the highest grade in the county.

As a result, he received a four-year scholarship at Macomb Academy, which was the new state normal school. He entered Macomb Academy in 1910, and received scholarships through the completion of his doctorate. He finished the Macomb Academy program in three years, then taught one year as principal of a two-year high school at Pleasantview. He then returned to Macomb Academy for two years and took every course offered in geography and geology. In 1916 he went to the University of Chicago and received a B.A. degree in 1917. He married on May 30 1917, then went to St. Cloud, Minnesota, as chairman of the Geography Department.

In February 1918 Clarence resigned and enlisted in the U.S. Army. Two weeks later, in Washington, D.C., he contracted measles and was sent to Walter Reed Hospital, where he also got chickenpox. Summer came and he was sent to New Jersey, where the Army examined his record and noted he had sold insurance while in college. The Army placed him in an insurance division where he remained until the armistice was signed on November 11 1918. He was then transferred to the infantry and was made a regimental sergeant major, the highest non-commissioned officer rank in the Army at that time. He was responsible for receiving troops returning from overseas.

Jones returned to the University of Chicago in September 1919 and studied there until he received his Ph.D. in September 1923. The departmental faculty then at the Univeristy of Chicago consisted of Rollin D. Salisbury, Harlan H. Barrows, Walter S. Tower, Charles C. Colby, Wellington D. Jones, J. Paul Goode, Ellen C. Semple, Robert S. Platt and Derwent S. Whittlesey. At that time the University of Chicago was the most influential and significant academic center of geography in America. Charles C. Colby had developed an expertise on ocean trade and transportation and taught a course on that for many years. He had numerous graduate students write dissertations on ports; Clarence wrote his doctoral dissertation on the port of Montreal, under Colby.

While he was a graduate student at Chicago, Jones shared an office with Richard Hartshorne. He also established contact with Alexander Bell, who owned a large and fashionable department store. This led to his tutoring Alexander Bell's son for entry into Harvard University. From that contact, Jones became a leader in boys' club after-school activities and Saturday field trips. Through this work, Jones met George A. Raney, secretary-treasurer of the International Harvester Company, who provided valuable support for Jones in Argentina during the summer of 1925. Jones made more money from his boys' club work than he did in his first year of teaching at Clark University ($3,000 in 1923).

Clarence Jones took all of Ellen Churchill Semple's courses at the University of Chicago. She helped him with his first article, which was on the Falkland Islands, and was published in the *Geographical Review*. She came often to the Jones house in Worcester and was a good friend of the family. Because of their close friendship and professional respect, Semple asked Jones to revise her first book, *American History and Its Geographic Conditions*. Jones's revised edition was published in 1933.

Jones went to Washington, D.C. in September 1942 as head of the South American Division of the Board of Economic Warfare. This division was primarily responsible for securing raw materials for World War II. Later, in Washington, he worked for the National Planning Association along with Jorge Zarur, a Brazilian geographer and close friend of Cotton Mather, who had introduced Zarur to Jones. These three geographers were close associates on Latin American matters during the War. In addition, Jones had a Peruvian geographer, a geographer-economist from Argentina and a specialist on Mexico on the Board of Economic Warfare. Later during the War, Jones worked in the Intelligence Agency, the Office of Strategic Services.

An important geographic contribution of Jones was a land classification project in Puerto Rico, under the auspices of the Puerto Rico Planning Board, which was headed by one of his former graduate students, Rafael Pico. Jones, with John Lounsbury as his assistant, used 1:10,000 scale air photos and the Vernor C. Finch "fractional code method" to map Puerto Rico's land quality and land use between 1945 and 1953. The Puerto Rico survey demonstrated the value of a detailed resource inventory in economic planning.

Clarence F. Jones and Preston E. James were the two leading twentieth century American scholars on the geography of Latin America. They cooperated closely on numerous projects, coedited *American Geography: Inventory and Prospect,* and had a long association with the Pan American Institute of Geography and History.

Jones wrote many articles, mostly on Latin America. He authored a classic textbook on the geography of South America and a volume entitled *The Commerce of South America*. With Gordon Darkenwald, one of his former students, he coauthored a renowned text on *Economic Geography*. He was president of the Association of American Geographers in 1957.

Clarence F. Jones, affectionately referred to by his many graduate students as "Pappy," was one of the most important mentors of graduate students in America in the twentieth century. During his career, he was the advisor of about 200 graduate students for the M.A. degree. Over fifty students received Ph.D.'s under his supervision at Clark University (1923-42) and at Northwestern University from 1945 until his retirement on August 30 1961.

Three years later Jones moved to the family farm near Oneida, Illinois. In 1965 he finished the last revision of *Economic Geography*. Most of Jones retirement was spent on the "Home Place" — 160 acres of fertile Illinois farmland. He was an astute manager; by 1985 the family estate had expanded to 480 acres. At ninety-two, he was still involved in the farm's management, and drove daily to Galesburg for lunch. His death in late 1991 at ninety-eight ended the career of one of America's most remarkable leaders in geographic education.

Armin K. Lobeck

Armin Kohl Lobeck
August 16, 1886 - April 26, 1958

Armin Kohl Lobeck was born in New York City. When he was only three years old his family moved to Haworth, in northeastern New Jersey. The eldest of five sons, he withdrew from school to help support the family when his father became ill. He was soon able to return to high school and complete his secondary education in Haworth.

At twenty-one, and older than most high school graduates when he entered Columbia College in 1907, Lobeck tried to make the most of his educational opportunity. He continued to live in Haworth and made the laborious round trip to the University each day to study architecture and botany. He graduated in 1911 with an A.B. degree. In his senior year at Columbia he enrolled in Graduate School and was awarded a masters degree in 1913 (Smith, 1959:83-87).

In 1911 Lobeck began his teaching career at the Philadelphia College of Pharmacy, where he taught botany from 1911 to 1914. Returning to Columbia to continue graduate work in physiography and geology, he came under the influence of Douglas W. Johnson, an internationally known scholar in the field of geomorphology.

While studying structural geology and geomorphology, Lobeck began to experiment with and apply his knowledge of architecture to the representation of land forms, a method which became known as physiographic drawing. After spending several months in Puerto Rico engaged in field work for his dissertation, Lobeck returned to Columbia and received his Ph.D. in 1917.

When World War I ended, Lobeck became a member of the Geography Section of the American Commission to Negotiate Peace, and in December 1918 he went to France. In Europe he was closely associated with nationally known American geographers such as Isaiah Bowman, Mark Jefferson, Lawrence Martin, and others. Lobeck's duties consisted of preparing physiographic maps of problem areas such as the Balkans and Albania. After the members of the American Commission returned home, these maps were published by the American Geographical Society.

It was at the Peace Conference in Paris that Lobeck met Professor C.K. Leith, Chairman of the Department of Geology at the University of Wisconsin. Professor Leith was looking for a physiographer-geographer who could fill a position in his department that had been vacant for two years (since Lawrence Martin's resignation). Lobeck accepted the position with the rank of assistant professor and moved to Madison with his wife in the autumn of 1919, where he remained for ten years. During this time he prepared and published many physiographic maps.

Lobeck's large-scale "Physiographic Diagram of the United States" was published by A.J Nystrom. Lobeck immediately began work on a small-scale edition of this map. The smaller map, with its accompanying text, was published in 1922 by the Wisconsin Geographical Press, which was wholly owned by Lobeck. His publishing business was very successful. He took the business with him to Columbia in the autumn of 1929 when he returned there as a professor.

By the time he retired in 1954, after a quarter of a century of teaching, hundreds of undergraduates and graduate students had heard his lectures and were amazed and fascinated by his blackboard sketches, which made land forms come to life. Block diagrams and other sketches, as well as well-organized lectures, made him a popular and effective teacher.

For his distinguished service as a teacher of geology and geomorphology, Lobeck was awarded the Neil Miner Medal by the Association of Geology Teachers in 1958. Geography teachers are still using the class and laboratory-sized exercises of the *Physiographic Diagram* of the United States and *Physiography Types,* which were originally published by the Wisconsin Geographical Press. In 1954 the C.S. Hammond Company began to publish Lobeck's physiographic diagrams.

Lobeck is best known among geomorphologists, geologists, and geographers for the wide variety of his books, reports, and exercises that contain one or many physiographic maps which were prepared by him or by his students.

Lobeck was employed during World War II by the Military Intelligence Service, General Staff of the U.S. Army (Smith, 1959:85). His maps were used for the invasion of North Africa. Later he prepared a large-scale physiographic map of Europe, which was prepared in sections to meet the needs of the General Staff for planning operations on the continent. After all the sections were finished they were pieced together to become a single physiographic map of Europe.

Whether teaching in the classroom, working for the government, or engaged in research, Lobeck displayed many characteristics of the early naturalists and geographers such as Nathaniel Southgate Shaler. His early interest and training in and later teaching of botany were valuable assets when he served on the Board of Trustees of the National Parks Association and on the Board of Governors of the Nature of Conservancy.

After retirement Lobeck and his wife spent their summers at Martha's Vineyard so he could rest and refresh himself. However, he continued to work on publications and maps, scheduling a definite number of hours each day for work. He finished *Things Maps Don't Tell Us* in 1958. Even when his activities were restricted by a serious illness, he continued to make hourly additions to a manuscript.

Smith characterized Lobeck as follows:

> His personal life was characterized by devotion to his family and friends. He had a warm-hearted personality and a rare sense of humor cherished by all who knew him. He was a genuinely magnanimous person and was amused at those with fewer talents and more modest skills who were always pressing for recognition (Smith, 1959:85).

Almon Ernest Parkins

Almon Ernest Parkins
January 10, 1879 - January 3, 1940

Professor J.R. Whitaker, who succeeded Almon Parkins as professor of geography and head of the Department of Geography at George Peabody College for Teachers, Nashville, Tennessee, characterized Parkins as follows:

> Ernest Parkins will be remembered as one of the most American of American geographers. In his training, in his teaching, in the results of an exceptionally productive career, in his friendship, he stands as the product and representative of contrasted sections and of the entire nation (Whitaker, 1940:167-168).

Professor Parkins was born in Marysville, New York. In 1906 he graduated from Michigan State Normal College, Ypsilanti, Michigan. It was the remarkable ability of Mark Jefferson to make the study of geography such a high adventure that influenced Parkins to decide on a career in geography. Jefferson guided Parkins to the University of Chicago, where he studied with many of the teachers who molded American geography. Here Parkins came under the influence of Rollin Salisbury and H.H. Barrows (Whitaker, 1941:46-50). Parkins received his doctorate from the University of Chicago in 1914.

Parkins's earlier writings reflect the years of firsthand study in the north and his lifelong concern with the teaching of geography. Between 1910 and 1930 he published at least twelve articles in the *Journal of Geography*. In addition, other articles were published in the *Bulletin of the American Geographical Society* and the *Bulletin of the Geographical Society of Philadelphia*. The 1920 *Journal of Geography* masthead indicates that the year George J. Miller became editor, Parkins was one of four associate editors (Barton, 1964:345-355). Parkins had been contributing editor of the *Journal of Geography: A Magazine for Teachers* the previous year, when the magazine was edited by Isaiah Bowman, director of the American Geographical Society.

Parkins was appointed instructor of agricultural geography at the University of Missouri in 1914. Two years later he was named professor of geography at George Peabody College. Within a year he was a member of both the National Council for Geography Teachers Executive Committee and its Board of Directors. By remaining at Peabody until his death in 1940, Parkins exerted a lasting, constructive influence on teachers and geographers of the South and on geography in general through his teaching and writing.

One of Parkinss' most distinguished proteges wrote:

> He sought to uphold the traditions of a graduate institution through his own activites

Robert S. Platt

in original scholarship, and of a college of education by his attention to the educational aspects of geography. His writings reveal the desire to serve at all points in the propagation of the best of geographic truth, from the discovery of basic knowledge of human significance to the dissemination of that knowledge. In editing the book on The Teaching of Geography, published by the National Society for the Study of Education, he helped to strengthen the position of Geography in the public schools at a relatively critical time (Whitaker, 1940:167-168).

In addition to authoring *The South*, Parkins was joint author of the McMurry-Parkins geographies and of the Miller-Parkins textbook on North America (1928, 2nd edition, 1934), and was coeditor with J.R. Whitaker of *Our Natural Resources and Their Conservation* (1936, 2nd edition, 1939). In 1934 the National Council of Geography Teachers Distinguished Service Award was conferred on Parkins.

Robert Swanton Platt
December 4, 1891 - March 1, 1964

Robert S. Platt was born into a well-to-do family in Columbus, Ohio (Hartshorne, 1964:630-637). He attended elementary schools in Columbus, and St. George's and Hotchkiss schools in New England. At Yale he majored in philosophy and minored in history. He was elected to Phi Beta Kappa and graduated in 1914. He spent a year at Yale-in-China, where his duties included teaching geography. Afterwards he enrolled in the Department of Geography, University of Chicago, as a graduate student.

Platt had a great curiosity about the world as a whole and the apparent instincts of an explorer. Undecided about a career after graduating from Yale, he thought he would try a year in China without knowing what he would study. After a stay in Scandinavia, he planned to travel to China by crossing Russia to the Near East, joining some other Yale students in India, and then proceeding to his destination. As a result of events in 1914 he was detained in Russia as a spy. After he was freed, he finished crossing the Eurasian continent via the Trans-Siberian Railroad, which proved more adventurous than planned. At long last, after arriving in China, he was assigned to teach geography with little preparation other than a geology course at Yale.

A third of a century later, reflecting on his first teaching experience in China, he remembered his geography class of Chinese students on trips into the countryside of Hunan Province.

As Platt returned to the U. S. from a trip around the world, a second fortuitous situation occurred. A traveling companion, having inquired about Platt's activities as a teacher in China, said:

> My brother-in-law, Wellington Jones, who is teaching at the University of Chicago, is interested in much the same sort of thing. He calls it all geography (Hartshorne, 1964:631).

Platt proceeded directly to Chicago to investigate geography as a university subject and as a possible career. After enrolling in the geography graduate program, he was appointed an instructor in 1919 and received his Ph.D. in 1920. He advanced to full professor in 1939 and became chairman ten years later. After eight years as chairman, he became professor emeritus in 1957. In later years, Platt remarked that he stayed at the university almost fifty years "to study geography."

In 1922 Platt married Harriet Shanks, who had been a student in his classes and was also an adventurous, spirited young woman. Harriet joined him in planning and operating the summer field camps whenever possible. Field studies were made at a variety of different locations in the Upper Great Lakes region — as far north as James Bay, and west and east along the international boundary.

His second area of research during the first half of his career was Latin America. Prior to World War II he made seven trips to that continent, each one from two to six month's duration. After the first trip, his wife accompanied him, assisting in the field work by conducting interviews in Spanish. In the early years they traveled by boat and train; later, in the 1930s, they traveled by plane. Over the years they made spot studies in rural areas in most of the major regions of each of the Latin American countries.

During World War II Platt served his country in the geographic sections of different federal agencies. He served as adviser in the Office of the Geographer in the Department of State in 1943, and in 1944-45 was chief of the Division of Maps of the Library of Congress.

After the War, during 1947-48, the Platts made two additional field trips to South America. During the first of these, in 1947, he served as consultant for the Conselho Nacional de Geografia of Brazil, studying proposed sites for the new federal capital. During the 1948 trip he concentrated on the problems of Tierra del Fuego.

Professor Platt officially retired from teaching at the University of Chicago at sixty-five, but he continued to teach part of each year, either in a seminar at Chicago or as a visiting professor at Michigan State University, Rollins College, Ohio State University, and Indiana University.

Either in the classroom or in presenting papers at a meeting or in publications, Professor Platt was constantly teaching. He never took the

Douglas C. Ridgley
March 18, 1868 - October 10, 1952

Douglas C. Ridgley was born in Wabash County, Indiana. He graduated from Indiana State Teachers College, Terre Haute, in 1891. He received an M.S. from the University of Chicago in 1922 and his Ph.D. from Clark University in 1925.

Ridgley began teaching in the Indiana public schools before he finished his first degree. He moved to Chicago in 1895, where he taught at West Division High School for five years. He then was principal of the Victor F. Lawson Grammar School for three years (Burnham, 1953:82).

Dr. Ridgley moved next to Illinois State Normal University, where he was professor of geography and geology from 1903 to 1922. He then went to Clark University as a special lecturer and director of the Home Study Department. In 1924 he was appointed associate professor of geography in education, and in 1927 became a full professor. In 1934 he became professor emeritus.

Throughout his active teaching career and into his retirement, Ridgley accepted special assignments. In World War I he served in France as Director of Geography, Army Educational Corps, from December 1918 to July 1919. In World War II he was a lecturer in geography for the U.S. Air Force. He was professor of geography in the first "College Cruise Around the World" in 1926-27. He pioneered in organizing geographical excursions under trained leaders, a practice which at the time attracted much attention in the world of education. He was professor of geography and chairman of the Geography Department's aviation program at Jefferson College, St. Louis, Missouri, from 1943 to 1944.

Professor Ridgley was very active in the early years of the National Council of Geography Teachers. He was treasurer from 1921 to 1930 (Barton, 1964:345-355), served two three-year terms on the Executive Committee, and was president in 1931. Editor Miller appointed Ridgley as Associate Editor of the *Journal of Geography* in 1928. During the first two decades of the NCGT, Ridgley published thirteen major articles in the *Journal*. He received the National Council of Geography Teachers Distinguished Service Award in 1935. In addition to teaching, he was the author and coauthor of many textbooks, workbooks, and pamphlets.

Professor Ridgley was a pioneer in the use of visual aids in teaching geography. He equipped the geography and geology department at Illinois State Normal University with a large collection of flat and mounted wall maps, globes, and mineral and rock specimens. His major contribution to visual education was working with the Keystone Company in preparing geographic slides for stereopticon news; later they made geographic slides for classroom use (Marshall, 1956:276).

Rollin D. Salisbury

Rollin D. Salisbury
August 17, 1858 - August 15, 1922

Rollin D. Salisbury was only sixty-four years old when he died, but he had such a brilliant mind, such a clear vision of his profession and of the world, and so much drive that he accomplished more than seems possible during his forty-year professional career. In 1903 he organized, supervised, and headed the development of the first sizable department of geography in any American university. During the first sixteen critical years, Salisbury selected and built a distinguished staff who contributed notedly to the science of geography and to the training of numerous students who subsequently held prominent positions throughout the United States. The success of the department, under Salisbury's leadership, established geography as a separate department at the University of Chicago and influenced the establishment of geography departments in other universities.

Rollin D. Salisbury was born on a farm near Spring Prairie in Walworth County, Wisconsin. His parents were New Englanders, and his mother was a relative of William Cullen Bryant.

Salisbury attended Whitewater Normal School, where his uncle was president. He taught in a rural school for a year, and at the age of twenty went to Beloit College, where he worked for his board and room in the home of T.C. Chamberlain. After Salisbury graduated in 1881, he started working for the U.S. Geological Survey. In 1883, after Chamberlain left Beloit College, Salisbury became a faculty member at Beloit. He studied in German universities during 1887-88 and gained recognition by tracing a hitherto unrecognized moraine across northern Germany.

He continued to work for the U.S. Geological Survey. With Chamberlain, he wrote voluminous reports on the Driftless Area of Wisconsin. He participated in and/or supervised studies of glaciation in the western mountain ranges such as the Big Horn. In addition to being an exceedingly successful administrator, Salisbury's publications substantiated his research abilities. He contributed to the understanding of physical geography with his penetrating field studies in various regions, notably Wisconsin, New Jersey, Wyoming and Greenland, with lesser studies in Germany, South America, the West Indies, and other areas.

His research was not an end in itself. Deductions from his field studies were made available to teachers and students in books and textbooks. *The Physical Geography of New Jersey* appeared in 1898. In 1904-06 a three-volume treatise, coauthored with T.C. Chamberlain, appeared. Salisbury authored most of volume one, physical geology, and made major contributions to volumes two and three. The treatise was condensed into a widely-used *College Geology*.

In 1912 Salisbury published *Elements of Geography*, a widely used college text for more than a decade, and a year later an adaptation of *Elements of Geography* appeared in the high school text *Modern Geography*. Most of Salisbury's publications were in geology. In a memorial in the *Bulletin of the Geological Society of America* (March 1931), there is a bibliography of nearly one hundred items, and in the *Annals* (March 1953) there is a list of his publications. Salisbury's physical geography publications helped bring recognition to geography as a physical science and helped provide a sound physical basis for the development of the human phases of the subject.

Salisbury was not only a nationally recognized researcher and author, but was also a superb teacher, lecturer, and editor:

> He was so successful as a teacher at Beloit College, where he taught for seven years, that when he was offered a professorship at the University of Wisconsin, a petition signed by practically every Beloit student implored him to remain, which he did for another year. At Chicago, to which he went upon its opening in 1892, many students rated him as their most successful teacher (Visher, 1953:6).

Salisbury was in demand as a lecturer; he spoke frequently, especially to groups of teachers. His lectures were not frivolous and he did not speak down to his audiences. His lectures were popular because of their high educational effectiveness.

A significant part of Salisbury's teaching was his earnest, consistent effort to find and help develop the promising students who were enrolled in his classes. He selected from each of his classes a few whom he considered especially gifted, and devoted much of the class period to increasing their comprehensiveness and effectiveness. Occasionally, he concentrated on one or two of these students for several consecutive days. Despite considerable gruffness and brusqueness in the classroom, these special students overlooked his style and idiosyncrasies because of his drive and earnest desire to have them learn.

Salisbury stressed the need for daily preparation, and followed his own philosophy by being well prepared for each of his classes. He encouraged his students by emphasizing:

> One of the great lessons which the world needs to learn is that progress comes from cumulative achievement. If every individual could be made to realize that even his tiny contribution to the sum of useful work is really moving the world along, it would add grandeur to life and dignity to all human endeavor. This is the frame of mind that should be developed in every person and cultivated until it becomes a habit (Visher, 1953:7).

One of his students, Visher, reported that Salisbury did almost no research in his later years, but devoted his energies to teaching and administration. His students made major contributions to the geography profession. Visher, writing in 1953, stressed the contributions Salisbury's students had made to geography. Ten of his students had already been president of the Association of American Geographers. Ten of the first nineteen recipients of the National Council of Geography Teachers Distinguished Service Award were students of Salisbury (Allison E. Aitcheson, W.W. Atwood, C.C. Colby, V.C. Finch, Alice Foster, George J. Miller, Edith Parker, A.E. Parkins, Zoe A. Thralls, and S.S. Visher).

Salisbury served geography with distinction and vital influence as an administrator, maintaining the high ideals of the department and selecting a staff with care and insight. With only one exception, his staff members became distinguished geographers. He screened assistants and fellows with such meticulous concern that they contributed to the strength of the department, and were so successful in their profession that they brought honor to their graduate departments.

> As dean of the graduate school of science, he contributed significantly to the establishment of policies which helped it become a world leader, and he helped geography attain parity among the university's departments (Visher, 1953:9).

As an innovator he was co-founder and first president of the Chicago Geographical Society, and promoted its publications.

After serving sixteen years with distinction and building a highly respected department, Salisbury relinquished the chairmanship of the department. His successor, Harlan H. Barrows, who was partly trained under Salisbury, helped establish geography as a social science at Chicago.

Because Salisbury performed his administrative duties efficiently and effectively, his responsibilities were increased. After serving as dean of the University of Chicago's Science College (1874-97), he became acting dean of the Graduate School of Science (1897-99) and dean from 1899 to 1922. He was also head of the department of geography (1903-19) and, following Chamberlain's retirement, became head of the department of geology (1919-22). In spite of a heavy administrative load, he accepted the presidency of the Association of American Geographers in 1912. He also served as managing editor of the *Journal of Geology* for nearly thirty years while he served simultaneously in other administrative capacities. Salisbury freed Chamberlain from many duties so that he had much more time for fundamental research.

Rollin D. Salisbury never married. He had a coronary thrombosis on May 31 1922 and died two and a half months later.

Carl Ortwin Sauer

Carl Ortwin Sauer
December 24, 1889 - July 18, 1975

Sauer's ancestors were members of a German pietistic sect that — under the designation of "German Methodists" — became part of the general Methodist organization. His father was a teacher at Central Wesleyan College, Warrenton, Missouri, where Carl was born. Sauer's grandfather was a minister of the sect.

Carl Sauer's education started at home. His family heritage was one of:

> . . . simple piety, love of music, respect for human learning, and by the standards of the American Middle West, of permanent residence on native soil. His heritage from the wider community into which he was born was of "the native values of rural life," industry, thrift, farsighted care for the land and for moveable possessions (Leighly, 1978:99-108).

Family connections with their fatherland were sufficiently close that when Carl Sauer was nine years old his parents placed him in a school in Calw, a small town in Wurttemburg. Five years of rigorous instruction there, with an emphasis on the classical languages, gave him a better basic education than he could have obtained in Warrenton. It enabled him, after he returned home, to complete his work for the bachelors degree at Central Wesleyan College in 1908. He read widely in the college library, but found his greatest intellectual stimulation in the publications of the U.S. Geological Survey and the State Geological Survey.

Primarily interested in geology, he first enrolled for graduate study at Northwestern University, Evanston, Illinois. He quickly found that petrology was emphasized in the Geology Department rather than those aspects of geology which would interpret the earth as he observed it around him. When he learned of the instruction in geography being given by Rollin D. Salisbury at the University of Chicago, he transferred there in 1909. Here Sauer appears to have been influenced by two professors — Salisbury, who headed the Department of Geography but whose courses were listed in the university catalogue under geology; and Henry C. Cowles (1869-1939), a pioneer of plant ecology in the United States whose field excursions led students to interpret vegetation as Salisbury's had led them to interpret land forms.

As was customary in the Geography Department at the University of Chicago, Sauer wrote his doctoral dissertation on a region — the Ozark Highland of Missouri — to the south across the Missouri River from his native Warrenton.

Sauer served as assistant geologist for the Illinois Geological Survey from 1910 to 1912. He worked for a year in Chicago as map editor for Rand McNally and taught for a year as an instructor at the State Normal

School in Salem, Massachusetts (1913-14). After receiving his doctoral degree in 1915, he became an instructor in geography at the University of Michigan, Ann Arbor, where he remained for seven years, becoming a professor in 1922.

While at the University of Michigan, Sauer and some of his colleagues became interested in the problem of cut-over and burned-over lands of the northern part of lower Michigan which were reverting to the state for non-payment of taxes. Sauer and others persuaded the state to undertake a "Land Economic Survey" of the qualities and potential utility of the cut-over lands. Work began on the survey in 1922 by the experimental mapping of one county. In 1923 Sauer accepted an appointment as professor of geography at the University of California, Berkeley, and ended his connection with the survey.

At Berkeley, Sauer took charge of a tiny department staffed by himself and two associates (Leighly, 1978:100), who were not yet qualified by a doctorate for a regular appointment. The department had previously been little more than an appendage of the much larger and nationally prominent geology department. Immediately upon arrival, Sauer had to appeal to the University administration since the geologists had encroached upon the limited area available for geography.

Faced with the problem of reorganizing the department, he initiated a new sequence of courses. The introductory course became physical geography, built around the world distribution of climatic regions; the second course was organized by cultural regions.

Finding German geography the best example to imitate, at different times he brought to the department Oskar Schmieder, Goddfried Pfeifer, Wolfgang Panzer, and Albrecht. Sauer also found new directions for his research. Before leaving Michigan:

> he had become dissatisfied with the formulation of the task of geography he had learned at Chicago — the rather abstract definition of the relations that subsist between human groups and their environments, the environment being conceived as the independent variable. He sought a more concrete formulation, one that would emphasize rather than abstract "relations." Soon after he arrived in Berkeley he attempted a program directed toward that end. The result of his efforts was The Morphology of Landscape (Leighly, 1978:100).

The Morphology of Landscape attracted the attention of academic geographers in the United States, and it became recommended or required reading for graduate geography students, but exerted little influence on the substantive work that Sauer and his students did after it was published. By 1925, when the essay appeared, Sauer had discovered Baja, California, as an easily accessible region where little scientific work had been done and where a dry winter season enabled roads to be passable during winter vacations. For many years Sauer and his students used the winter vacations for field work in Baja, California, northern Mexico, and adjoining parts of Arizona and New Mexico.

In 1927 Sauer produced *Geography of the Pennyroyal,* which was written in Berkeley and based on material he had taken west with him. It was his last publication in regional geography. Henceforth, his field work was directed toward the solution of problems, not toward the characterization of regions.

Field work raised a wide range of deep questions in Sauer's mind. Some of these questions concerned historic and prehistoric settlements (indicated by archeological remains), movement, occupation in Spanish time, and especially native agriculture and plant cultivation. These studies culminated in *Agricultural Origins and Dispersals (1952).*

In 1954 Sauer relinquished the chairmanship of the department, which he had retained longer than was customary at Berkeley. He had already begun work on his next enterprise — the organization of a symposium on "Man's Role in Changing the Face of the Earth," held at Princeton, New Jersey, June 1955. In the autumn of 1953 William L. Thomas, Jr., then assistant to the director of the Wenner-Gren Foundation of Anthropological Research, invited Sauer to act as chairman of the conference. Sauer gladly accepted, but proposed the conference be delayed until 1955 to gain time for thorough preparation. A combination of Thomas's organizational talents and Sauer's wisdom and fertile mind resulted in a highly successful symposium. Thomas edited the proceedings in a volume with the same title as that of the symposium.

At his death, Sauer left a fragmentary outline of his next undertaking — an historical geography of the United States titled "Recessional at the Bicentennial." His notes indicate that he planned to emphasize the prodigality with which his compatriots had squandered the native riches of their land, and the excessive use and waste that confronted them with what might have been the most critical problems yet encountered in their history (Leighly, 1964:101).

During his lifetime, Sauer demonstrated that he had a superior mind and was an original thinker. He received international recognition for his research and, in addition, was an excellent teacher. His students report that he authenticated the lecture system by presenting material unavailable in print and expressing unique points-of-view. He did not "talk down" to them.

Stanislawski writes:

> Probably Sauer's greatest contribution as a teacher came not by injunction, but by example of persistant curiosity and search for knowledge beyond present limits (Stanislawski, 1975:548-554).

Ellen Churchill Semple

Ellen Churchill Semple
January 8, 1863 - May 11, 1932

In 1897 Ellen Churchill Semple made her initial modest contribution to American geography with an article entitled "The Influence of the Appalachian Barrier Upon Colonial History," which was published in the *Journal of School Geography*. Her last work, *The Geography of the Mediterranean Region*, appeared in 1931, approximately thirty-five years later. During this time she gave distinguished service to geography, promoting it from coast to coast in the United States by her lectures, teaching, and publications. She not only became the most distinguished American woman geographer of the twentieth century but became the greatest anthropogeographer in the world (Colby, 1933:229-240).

In her early forties, Semple was selected as one of the forty-eight charter members of the Association of American Geographers. At the time she was selected she had never taught a college-level course in geography, nor had she earned an advanced degree in the subject.

Miss Semple was born in 1863 in the bustling, thriving city of Louisville, Kentucky, during the Civil War. She was born into a fine family and lived her girlhood in an atmosphere of wealth and refinement. Her father, Alexander Bonner Semple, had immigrated from Scotland and settled in Pittsburgh, then moved to Louisville to establish a wholesale hardware business. Her father was fifty-eight and her mother forty-one when Ellen was born, the youngest of five children. The Semples were among the socially prominent families who gave their children good schooling, an abundance of books, and a healthy, well-ordered life. Ellen, who was educated in a private school, was fascinated by books on travel and history, excelled in tennis, and was a skilled horsewoman. She enjoyed taking long rides out into the countryside alone so she could ponder over what she had been reading.

Semple's educational and scholarly life may be organized into three sections: (1) the early years before the publication of her first book; (2) between the publication of her first and second books; and (3) between the publication of her second and last book.

In 1878 Ellen entered Vassar, although she was only fifteen years old. Having attended a private girl's school in Louisville, where she was an excellent student, she passed the entrance examinations with ease. With good health, a cultivated bearing, and with a wealth of knowledge accumulated from extra reading, Semple attended Vassar, where vigorous training and stimulating experiences were added to her life. Here she associated with a cosmopolitan group of students who had come from different parts of the United States and the world. This experience broadened her social horizons. The training she received in English composition, the classics, history, and economics provided work habits which subsequently proved invaluable. In 1882, at the age of nineteen, she received her bachelors degree. She was elected to Phi Beta Kappa, and as valedictorian was selected to give the commencement address, which was entitled "The Conscience of Science."

Semple returned to Lousiville, where one report says she joined the social life of the city. However, the social whirl bored her so she began teaching in a private school operated by her sister. However, teaching did not sufficiently stimulate her mind either, so she studied sociology, economics, and fragments of geography. She then began writing her thesis for an M.A. degree in history at Vassar, which she received in 1891.

Either during the time of extensive reading she did in Louisville while teaching or while she was studying for her masters degree at Vassar, she came in contact with the writings of Friedrich Ratzel, whose work at Leipzig was attracting worldwide interest.

In the summer after she received her masters degree she was traveling with her mother in England when she met a young American student who had studied with Ratzel. The student reported enthusiastically about his work. Instead of returning to America with her family, Semple went to Leipzig to study with Ratzel. Women students at that time were almost unknown in German universities. She was not permitted to matriculate, but she heard Ratzel's lectures and attended his seminars. Her work with Ratzel proved to be the turning point in her life and career. Contact with his masterly mind quickened her interest and convinced her that geography was the discipline she had been seeking. She remained in Leipzig for a year and a half, where she studied with new interest and intensity. She returned to America for a time, but in 1895 she was drawn back to Leipzig for further study with Ratzel. This time she was welcomed as a fellow student in a chosen field (Colby, 1933:231).

Following the death of Ratzel in 1904, Semple wrote:

> His achievement was in anthropogeography which he was the first to raise to the rank of science (Colby, 1933:232).

Years later Charles C. Colby wrote:

> If Ratzel raised anthropogeography to a science, Semple illuminated its pages with work which has high artistic values as well as sound scientific qualities. She gave the science, moreover, needed content, and effectively promoted its acceptance as a field of inquiry (Colby, 1933:232).

Upon returning to America, Semple became active in research in her chosen field. From the time her first article appeared in 1897, publications of important dimensions averaging more than one per year appeared for the remainder of her life. Most were published in nationally-known magazines. Her first articles were notable primarily because they introduced anthropogeographical methods into journals for American students.

In 1901 the *Geographical Journal* published her study of the "Anglo-Saxons of the Kentucky Mountains," which is highly ranked among geographical articles. It demonstrated Semple's ability to work efficiently in the field, since the study was based on information gained by riding on horseback into a little known and inaccessible area. Two years later her first book, *American History and Its Geographic Condition*, appeared. This volume was widely read and critically discussed by professionals in other fields, and involved its author in several interesting and stimulating controversies. In the field of geography it won wide acceptance as a textbook for courses on the historical geography of the United States.

After the appearance of her first book, Ellen Semple became a person of importance who was in great demand. In 1906 Professor R.D. Salisbury invited her to be a visiting lecturer for one quarter in the newly established Department of Geography at the University of Chicago. Semple and her work were enthusiastically received, and she continued to lecture at Chicago practically every alternate year until 1924. Her teaching fired the enthusiasm of graduate students who were just beginning their geographic work.

Semple enjoyed university teaching but was reluctant to take so much time from her research and writing for it. From 1904 to 1911 inclusive, she published eleven articles and her second book. Two of these articles were based on field work; three were philosophical studies dealing with the interrelation of geography and history.

Her second book, *Influences of Geographic Environment*, which was published in 1911, proved to be another controversial volume. This book had been originally undertaken at the instigation of Ratzel, who — prior to his death in 1904 — asked Ellen Semple to translate his great work *Anthropogeographie* into English.

Semple found the request impractical for several reasons (Colby, 1933:233). Following some preliminary study, she decided to write an English volume along the general lines of Ratzel's book, but greatly modified it.

Following the publication of her second book, Ellen Semple took a trip around the world — visiting Japan, China, the Philippines, Java, and India. As a person, she received great pleasure from the experience; but Semple, the scholar, could not be left at home. Through the influence of a Japanese Vassar schoolmate, she was shown official courtesies not available to the general traveler. Semple was soon involved in detailed field work in Japan. She returned home to write two articles on Japan, one of which dealt with Japanese agriculture and was such an effective portrayal of both the static and dynamic aspects of the landscape that it was translated into Japanese and published in a Japanese magazine.

Returning home from her around-the-world trip she found that the publication of *Influences of Geographic Environment* had added materially to her reputation both at home and abroad. During the summer term of 1912, she was invited to lecture at Oxford University in England. Her work was so well received that she was invited back for the summer of 1914. Following her lectures at Oxford in 1912, she lectured before the Royal Geographical Society in London, presenting for the first time the results of her study of Japanese agriculture. Later, in November and December, she appeared before the Royal Geographical Society in Edinburgh, Glasgow, Aberdeen, and Dundee.

Following her return from Europe, Semple lectured at Wellesley College during 1914-15 and at the University of Colorado the summer of 1916.

Ellen Semple's first article on the Mediterranean was published in 1915. There is no evidence that her interest in the area began at an earlier date because in two of her earlier magazine articles she dealt at some length with the maritime activities of the Phoenicians, Greeks, and other peoples of the ancient Mediterranean. However, from 1915 on, the Mediterranean claimed most of her interest, time, and energy. She traveled in the area and worked in the great libraries of Rome, Paris and London, gathering materials for her last volume.

Semple's work was interrupted during World War I. In the fall of 1917 she gave a course of lectures on the geography of the Italian front to officers at Camp Zackary Taylor near Louisville, Kentucky. From December 1917 to December 1918 she conducted special studies on the Mediterranean region and Mesopotamia for the Bureau of Inquiry for the Peace Terms Commission. This research was housed in the American Geographical Society building on Upper Broadway in New York City. The nearness of this building to Columbia University made it easy for her to accept an invitation to lecture there during the summer of 1918.

After the war ended she again returned to research, writing and teaching at the University of Chicago. In 1921 Semple became a professor of anthropogeography at the Graduate School of Geography, Clark University, where she remained until her death.

Semple was invited to serve as a visiting professor in various parts of the United States. She gave a series of lectures at Vassar College and prized the recognition from her alma mater. She gave a special series of lectures at the University of Michigan, the University of Kentucky, and at other colleges and universities. She taught the second semester of 1925 at the University of California at Los Angeles. In 1914 the American Geographical Society awarded her its Cullum Gold Medal. She was elected president of the Association of American Geographers for 1921. Her presidential address, "The Influence of Geographic Conditions Upon Ancient Mediterranean Stock Raising," given at the annual meeting in Washington, D.C., December 1921, was later printed in the *Annals*.

Throughout the last decade of her life, Ellen Semple's thoughts and efforts were concentrated on her work in the Mediterranean; her interest and enthusiasm for the subject seemed to grow each year. She began to limit her outside lectures in order to conserve time and energy to complete her last book, *The Geography of the Mediterranean Region.*

Her strict schedule and careful plans were interrupted in the late autumn of 1929 when she was stricken with a serious heart attack, which kept her in the hospital until the following summer. As her illness stretched from weeks into months, her friends and colleagues were concerned for her life as well as her being able to finish her Mediterranean study.

Slowly she regained her strength and at last became active again, but by summer's end she could at most maintain only a schedule of two hours of work per day while confined to bed. In September she was moved to a home facing the Clark campus, but was still confined to and worked on her Mediterranean manuscript in bed. In addition she began a revision of *American History and Its Geographic Conditions* with the cooperation of her colleague, Clarence F. Jones. In the early summer of 1931 she moved to Petersham, Massachusetts. There, with the assistance of a former student in the documentation and editoral phases of the work, she slowly but finally completed the Mediterranean book. Work was also completed on the revision of *American History and Its Geographic Conditions.*

By October 1931 Semple was strong enough to start her migration south to warmer areas. She remained in Asheville, North Carolina, for a time and then moved to West Palm Beach, Florida, where she died in 1932.

Nathaniel Southgate Shaler

Nathaniel Southgate Shaler
1841 - 1906

Although Professor Shaler taught geology courses for over forty years (1864-1906) at Harvard University, his courses consisted mostly of geographical content. His three-volume work, *The United States of America,* has been recognized as the first real geography of our nation (Bacon, 1955:349-353). His book, *Man and Earth,* which contributed to the nation's inventory of natural resources and to an early assessment of man's demands on them, established Shaler as an early leader in the movement to conserve natural resources (Koelsch, 1979:133-139). J. Russell Smith, Columbia University, in the first edition of *North America,* stressed Shaler's stature as a geographer by characterizing him as "a pioneer appreciator of the relationship between earth and man." Shaler, a public lecturer and author of twenty-six books and over 200 articles, developed a formidable reputation as an inspiring if not rigorous teacher of undergraduates. He aided in the establishment of geography as a learned profession in American universities.

Nathaniel Southgate Shaler was born into a well-to-do family in 1841. He grew up with an interest in domestic and wild creatures and by the age of twelve had deep interests in history and geology.

In 1858 the seventeen-year old Shaler, realizing he was more of a naturalist than humanist, enrolled in Louis Agassiz's laboratory in Harvard's Lawrence Scientific School, where he worked for three years among a brilliant group of naturalists. He became unique among Agassiz's students; while most of them concentrated on zoology, he concentrated on geology. Although Agassiz had him work on paleontology problems, Shaler's interest in the earth's surface grew through his scientific associations in the Boston Society of Natural History and through field work with outstanding geologists in New England.

A turning point in his research and teaching career occurred in the summer of 1861 when Shaler, with two other students, went on a four-month expedition to the Gulf of St. Lawrence in search of Silurian fossils. But the most important result of the trip was that Shaler developed a lifelong interest in the zone of contact between the land and the sea.

Following his return to Cambridge after two years of military and civil service in Kentucky during the Civil War (he served as a Union artillery officer), Shaler assumed the responsibility for basic instruction in zoology and geology in the Scientific School from the failing hands of Agassiz. At the early age of twenty-eight, Shaler was appointed professor of paleontology (later of geology) at Harvard College.

Shaler served for seven years as director of the Kentucky Geological Survey and created several summer field schools of applied geology for his students (including William Morris Davis). He also worked for the U.S. Geological Survey, making many geological and geographical investigations along the Atlantic Seaboard. Intermittently he worked as a mining consultant and held a number of appointments on Massachusetts State Scientific Commissions.

Shaler's travels in search of health or intellectual stimulus took him to England and to most of the Western and Mediterranean European countries, where he came into contact with leading scientific men such as Charles Darwin (whose theories he adopted although they were contrary to Agassiz's teachings), T.H. Huxley, and Sir Francis Galton. Travel intensified his sense of the significance of the humanized landscape, a theme on which he later wrote and lectured in the Harvard School of Architecture.

In a series of articles and books between 1880-96, he wrote mostly about the mutual relationship of nature and man in America. One of his best known books is *Nature and Man in America.* In this volume and in other articles, Shaler explored geographic problems such as the eastern range of the bison, an appraisal of prairies, the "Great American Desert," and the concept of climatic optimum. His treatment of problems such as these have in many cases been superseded by more recent studies, but his lasting contribution is that he raised the issues before a wide audience at an early date.

Shaler's diversity of interests, the conscientious fulfillment of his university administrative roles, and his personal attention to his students prevented him from making contributions to human geography on the order of his geological studies or from founding a "school" of human and regional geography (Koelsch, 1979:135).

Professor Shaler, influenced by his love of the earth's surface, established a rural retreat on his beloved Martha's Vineyard in 1888. His dreams of and plans for a productive retirement there were shattered by an unexpected illness and death during his last semester at the Lawrence Scientific School.

In 1979, seven decades after his death, Koelsch evaluated Shaler's career:

> As one whose insights and suggested hypothese prefigured many of our current professional concerns and those of the educated public, we may expect that in the last quarter of the twentieth century he will come into his own as a trailblazer whose pathways are still worth climbing" (Koelsch, 1979:136).

J. Russell Smith

Joseph Russell Smith
February 3, 1874 - February 25, 1966

Joseph Russell Smith, who was born near Lincoln in the northern part of the Virginia Piedmont, was reared on a farm by a Quaker father who was interested in new techniques of farming. Educated in liberal Quaker schools and by association with neighbors and friends in the Quaker community at Swarthmore, Russell found it difficult to accept the pettiness of the outside world. His desire to improve the world led him to major in economics when he entered the Wharton School. It took him from 1893 to 1898 to earn his B.A. degree, since he also had to teach in Quaker schools to pay his expenses at the University of Pennsylvania (Starkey, 1967:198:202).

For two years Smith carried out investigations for the Isthmian Canal Commission before he went to Europe for a year to study at the University of Leipzig (Orchard, 1967:128-130). He spent the final weeks of his European stay studying trade and shipping conditions at major European ports. He returned to the University of Pennsylvania, aided by a fellowship in economics, and in 1903 received his Ph.D. degree; the subject of his dissertation was the organization of ocean commerce.

Professor Smith began his teaching career as an instructor in the Wharton School of Finance and Commerce of the University of Pennsylvania, subsequently becoming professor and head of the Department of Geography and Industry. Since there was a serious shortage of good geography texts, Smith began writing textbooks. His first book, *Industrial and Commercial Geography* (1913), was widely used, had major influence, and lasted through many printings and new editions. In addition, he wrote many magazine articles and books on ocean carriers, the iron and steel industry, and industrial management. The last edition of *Industrial and Commercial Geography* appeared in 1955. Because of changing world conditions, the book had to be drastically altered. One of his two coauthors was his son, Thomas C. Smith.

In 1913 the University of Pennsylvania sent Smith on a field trip to Europe and to North Africa to do research on tree-crop agriculture. This experience provided the material for a later book and the incentive for his increasing involvement with conservation.

Smith had served sixteen years at the University of Pennsylvania when he was asked by the newly-founded School of Business of Columbia University to establish a curriculum in geography. Accepting this challenge, he developed regional courses in geography and created his famous research seminar, which was quite popular with his students.

Smith was a controversial figure among his colleagues in the two disciplines of economics and geography because of his unorthodox and daring approach. He remained at Columbia University for twenty-five years, and during that time published two major works, *North America* (1925) and *Tree Crops* (1929). After the publication of *North America*, he took a year's leave of absence to travel throughout Asia and broaden his geographical comprehension.

Although he lived in large metropolitan areas and taught in large urban universities, Smith remained close to the land. When he accepted the chair in economic geography at Columbia Business College in 1919, he continued to live in Swarthmore, taking the early morning train to New York on Wednesday and returning late Friday afternoon. During the summer the family moved to a log cabin located below the crest of Virginia's Blue Ridge Mountains. Seated on the front porch, one could look over the farmland which Smith's ancestors had tilled and where Smith continued to experiment with tree crops.

One year, when he took a leave of absence from the University of Pennsylvania to study and write on the influence of World War I on shipping for the Carnegie Endowment for International Peace, Professor Smith was faced with three monumental career choices. He was invited to Columbia University to develop a Department of Geography; he was asked to go to Russia and assist Herbert Hoover with famine relief; and a Philadelphia publisher asked him to write a series of elementary school geographies, stressing human interest rather than merely cataloging places and products. Years later Smith explained why he decided to go to Columbia. He said that educating future American businessmen at Columbia would do more for world peace than working on stop-gap famine relief (Starkey, 1967:198-202).

However, after going to Columbia, administrative red tape and politics turned Smith's attention back to writing geography textbooks which would teach American children world understanding. In 1921 he began a vast program of publishing grade school and high school geography textbooks. With his customary enthusiasm, he assembled masses of photographs, diagrams, charts, maps, and some of his own research into textbooks which were provocative, imaginative, and well-tuned to student response. He was a genius at language, with the ability to take numerous facts and weave them into sentences which excited, interested, and stimulated creative thinking. Smith's textbooks, which were adopted by state boards of education across the country, went through many printings and new editions.

How could one man do so much in a lifetime? One of his former students explains it this way:

> Undoubtedly he was fortunate in having a group of Wharton students who have since had distinguished careers in business, government, and universities. Some of these worked for him as he had worked for Emory Johnson. Perhaps more important, in 1898, he had the good fortune to marry Henrietta Stewart, only daughter of a Quaker store keeper in Southern New Jersey. She accompanied him to Europe, shared his intellectual life, and often helped him put his thoughts into life, and often helped him put his thoughts into final shape. More disciplined in many ways than her husband, she kept the "Geography Shop" going and ran a household which at one period included three sons, two or three dachshunds, three servants, a secretary, and several assistants who researched current projects. With all this responsibility, she still found time to edit manuscripts, do background reading and see that projects met deadlines" (Starkey, 1967:200).

Although Professor Smith concentrated on writing textbooks, he also wrote many significant articles while at Columbia. In 1927 he wrote "Plan or Perish" and was awarded the Harmon Prize for stimulating constructive public opinion in social and industrial fields.

Considering that Professor Smith started in history and shifted to economics, his geography was largely self-taught as was his science. His farm background provided him with practical biological information and the perspective for studying land use problems. His work in conservation paralleled and amplified his work in geography. He was a crusader of conservation, beginning as early as 1895 with a practical experiment — planting Persian walnut seedlings on one and a half acres of mountain slopes to prevent soil erosion. About 1909 he began to write articles for agricultural and scientific magazines and journals. His greatest success in geography was as a generalist; his works contain some dubious theories but also many amazing foresights (Starkey, 1967:201).

Among the many honors Russell Smith received were a Doctor of Science degree from Columbia University (1929), a Doctor in Economic Science degree from the University of Pennsylvania (1957), and the Cullum Geographical Medal from the American Geographical Society in 1956 (Orchard, 1967:130). In 1940 he received the National Council of Geography Teachers Distinguished Service Award.

Griffith Taylor

Griffith Taylor
1880-1963

Griffith Taylor, the son of a mining engineer, was born at Walthamstow, England. The family moved from there to Siberia, and in 1893 they moved to Australia. Taylor worked for a considerable length of time on four of the continents; his travels, studies, and publications included all continents.

"He was, perhaps, most widely known and respected for his authoritative studies of Australian problems" (Rose, 1964:622).

Taylor's professional life can be divided into five periods. Publications of the first and early period extended roughly through World War I and were in keeping with his initial training in geology and engineering. These publications were primarily concerned with problems of Australia's physical geography, such as its coral reefs and climatic and meterological divisions. Some of his publications dealt with Australian geology and paleontology. In 1910, between post-graduate training in England and an assignment in Antarctica, Taylor made an intensive field survey of the proposed federal territory and prepared a relief model of the area. He even suggested the name of Canberra, which was adopted in 1913, because he thought there was some aborginal significance to it. Later he participated in the geological survey of that territory.

The second period of his professional life was influenced largely by his interest in glaciation. In 1907 he first studied the results of glaciation on Kosciusko. After having received his B.Sc. and B.M.E. degrees from the University of Sydney, he came to North America to study the glacial district of Yosemite. While enrolled at Cambridge (B.A., 1910), Taylor's vacation periods were devoted to field studies of Europe's glacial topography, some of which were under the guidance of William Morris Davis. From among some 7,000 applicants, Taylor was selected to accompany the second and ill-fated Scott Expedition to Antarctica, for which he was the senior geologist and leader of the western parties (Rose 1964:622).

Taylor, while enroute to Antarctica, detoured to study New Zealand's glaciers. His Antarctic and glacialogical interests did not end with Scott's expedition, but continued throughout his life. Later on he surveyed and published articles on other glaciated areas. He made the first contour map of Kosciusko (7,316 feet), where he requested that his ashes be spread after his death. He followed later expeditions closely and sometimes evaluated their work. He published six books during this period. In 1916 he received an Sc.D. from the University of Sydney; his thesis was on Antarctic geology.

Taylor was appointed to the science faculty at the University of Melbourne in 1916; two years later he became a lecturer in geography. He had given lectures on geography at Sydney as early as 1907. In 1920 the McCaughey chair of geography was established and Taylor was appointed Australia's first professor of geography and head of its first department.

Taylor's book, *Climatic Control of Australian Production,* started a controversy which reached a cresendo in the 1920s. A major part of the book and his lectures were directed to the diverse and complex problems of Australia's settlement. His studies indicated that the resources of large areas of the Australian continent were less conducive to effective human habitation than journalists and politicians proclaimed. They anticipated a population expansion of at least 100 million or as many as 200 million; Taylor thought that Australia's resource potential indicated that only 20 to 30 million could maintain a high standard of living.

Taylor expressed his convictions articulately and frequently, but they generated controversy. Some of his books were banned, and he was called unpatriotic or worse. His Sydney chair became increasingly uncomfortable, and in 1928 he accepted a professorship at the University of Chicago.

The third period started at Chicago, where he further developed his interests in race, early migrations, environment, settlement, and history. His continued research resulted in his monumental work, *Environment and Race: A Study of Evolution, Migration, Settlement, and Status of the Races of Man.* Some authorities believe this book was "his most original contribution to the study of geographic distributions" (Rose, 1964:623).

However, Taylor soon discovered that the University of Chicago had one requirement that he objected to — forced retirement at age sixty-five.

The fourth period of Taylor's professional work started in 1935 when he accepted an appointment at Toronto, Ontario, Canada, to establish the first Canadian department of geography. He remained here until 1951; he studied, traveled, carried on field studies, and wrote on Canadian problems. During this period he produced ten major books.

The fifth productive period of his life started after his retirement from Toronto. Taylor moved to a pleasant suburb of Sydney in 1952. He started research on a new book, *Sydneyside Scenery and How It Came About.* Used as a guidebook, it sold quite well. In 1961 he collected 20,000 letters for the Commonwealth Archives and found pleasure in reliving arguments with many of the leading scientists of the U. S., Britain, and Canada in branches of research such as geography, climatology, glaciology, and anthropology. Near the end of his life a new Griffith Taylor Building for Geography at the University of Sydney testified to the love and respect his adopted country had for him.

John K. Rose characterized Griffith Taylor:

> Griffith Taylor was perhaps the most widely traveled and experienced geographer of his generation and possibly the most controversial person the field of geography has produced. . . . an unusual man, not overpowering in physique but with "presence," his high degree of energy and vigor, both physical and mental, continued little diminished into later life. In the classroom he was wide-ranging, with a rapid presentation, generally integrated with numerous slides made from his own freehand, three-dimensional block diagrams, and "mantle maps." Many of his scientific ideas were interdisciplinary and controversial; he apparently enjoyed scientific controversy (Rose, 1964:624).

One of the best but incomplete bibliographies of Taylor was first published in *Australian Geographical Studies,* Volume 2 (April 1964), pages 3-9.

Ray Hughes Whitbeck

Ray Hughes Whitbeck
April 2, 1871 - July 27, 1939

Ray Whitbeck was born in Rochester, New York. His youth was spent in the hilly upland of southern New York where his father was a village merchant. Professor Vernor C. Finch, head of the Geography Department, University of Wisconsin, and coauthor with Whitbeck, wrote:

> He well knew the restrictions that limited financial means can impose on ambition. From this background he carried into later years the modesty and friendliness, the direct approach to all matters of policy, and the severe simplicity in taste and manner of life that were to characterize him to the end (Finch, 1939:252).

Early in his career, Whitbeck became interested in teachers' problems. At the age of twenty-one he finished a training course at New York Normal School, Geneseo. He was a teacher and principal of Pike Seminary in New York for four years and Schylerville High School for two years. He obtained an A.B. degree at Cornell, where in 1901, as a departmental assistant, he came under the influence of the noted physiographer, R. H. Tarr. Financially unable to continue work in his chosen field of geography, after a year as assistant in physical geography at Cornell, he became a supervisor in the State Model School at Trenton, New Jersey, a position he held for eight years. During three of those years he also lectured at Adelphia College, Brooklyn (Williams, 1940:211-217). Whitbeck was also elected to membership in the Association of American Geographers.

At Cornell Professor Whitbeck not only acquired a strong background in physical geography but also developed a broad interest in human geography, including its political, social and economic phases, which had a great influence on his later writings.

In 1909 a turning point gave direction to the remainder of Whitbeck's life. He was invited to the University of Wisconsin to develop the human aspects of geography in a department where the physical side of the field already was strongly represented. This appointment gave him greater freedom to study and write; his scholarly articles, bulletins, and high school and college textbooks were widely accepted.

A year after joining the University of Wisconsin, Whitbeck became editor of *The Journal of Geography,* a post which he held until 1919. After resigning as editor, he continued as associate editor until his death in 1939. Throughout the early years of the National Council for Geography Teachers, Whitbeck gave unqualified support to both the organization and to George J. Miller, who became editor of *The Journal of Geography* in 1920 (Barton, 1964:345-355). Whitbeck took an active part in promoting the NCGT and accepted its presidency in 1920. He received the second Distinguished Service Award of the Council in 1933.

Professor Frank Williams, University of Pennsylvania, who was Professor Whitbeck's assistant during the first summer he taught at the University of Wisconsin, writes:

> As a teacher he was superb. His sincerity, personal charm, his interest in his students, his early training, and his careful preparation made him one of the most liked of professors on the various campuses where his services were represented. He regarded it his duty to present his work as clearly as possible and to select only those topics that he believed to be of value (Williams, 1940:213).

Professor Vernor Finch, who was Whitbeck's closest colleague at the University of Wisconsin, characterized the latter's writing and interest in students thus:

> He was led inevitably to think and write as much for the teacher as for the broadening of scholarship and the accumulation of knowledge in his chosen field. . . . One of his latest topics of conversation concerned the teachers of geography among his former students, inquiry as to their progress and hopes for their success (Finch, 1939:213).

After years of ill health, Professor Whitbeck was forced to retire in 1936. The Distinguished Service Award of the National Council of Geography Teachers which he received gave him great satisfaction during his last years. In addition to this honor he was also president of the Association of American Geographers in 1925.

Part 2 The Synthesizers

Oliver Edwin Baker

Oliver Edwin Baker
September 10, 1883 - 1949

Oliver Edwin Baker was born in Tiffin, Ohio. He attended the local public schools and, at nineteen, graduated from Heidelberg College in Tiffin with majors in history and mathematics. A year later he was awarded a masters degree in philosophy and sociology from Heidelberg. The following year he received a masters degree in political science from Columbia University. He spent a year at the Yale University School of Forestry (1907-08), and in 1908-12 did graduate work in agriculture at the University of Wisconsin, Madison. While there he also coauthored a Wisconsin Experiment Station Bulletin on the climate of Wisconsin and its effects on agriculture (1912). During three summers (1910-12) he worked with the Wisconsin Soil Survey. He was employed by the U. S. Department of Agriculture in 1912 and rose through the ranks rapidly, partly because of his active participation in the *Yearbooks,* to which he contributed for twenty-three years (1915-38).

As coauthor of the *Atlas of World Agriculture,* Baker received national recognition. In 1917, soon after the Atlas appeared, he returned to the University of Wisconsin for further graduate study. This time he majored in economics, receiving his Ph.D. in 1921. In 1937 he was awarded an honorary doctorate from Gottingen University, which is widely known for research in agriculture (Visher and Hu, 1950:328-336). The success of the *Atlas of World Agriculture* stimulated progress on the *Atlas of American Agriculture,* which he planned and edited. Portions of the Atlas were published between 1918 and 1936.

Baker was part-time professor at Clark University from 1923 to 1927. He was associate editor when *Economic Geography* first appeared; he later wrote a series of articles on the agricultural geography of North America for the journal.

According to Professor Visher, Baker became deeply interested in population issues about 1920. Baker realized, far more than did the average citizen, the great significance of wise leadership. He believed that leaders should have superior training, opportunities, and ability. He became deeply concerned over the fact that many of the more able people of America reared so few children (Visher and Hu, 1950:229). He maintained that for decades our exceptionally well-endowed people reared too few children to maintain their proportion in the population. Consequently, he lectured frequently to groups of well-educated people on the desirability of having more children.

In 1942 Baker was invited to establish a department of geography at the University of Maryland. When he retired from the chairmanship of the department in 1949 he had assembled an outstanding staff and had attracted a sizeable share of the University's undergraduate and graduate students. Before Baker died there were forty graduate students in the department.

Although Baker resigned as head of the department to devote himself to research and other projects, he continued to teach a relatively heavy schedule and gave generously of his time to students, visitors, and the staff. He initiated and assisted in two ambitious research projects — the *Atlas of World Natural Resources* and the *China Atlas*. Unfortunately, for the geography profession and world society, he did not live to contribute the best of his talents and skills to these atlases.

In Baker's early life his goals appeared to be simply to describe and interpret conditions of the immediate past. In his later years he endeavored to predict the future, assuming a continuation of various existing trends; he endeavored almost passionately to reverse undesirable trends. Throughout his life he believed and practiced the philosophy that geography could help improve society and also solve some of its problems.

Vernor C. Finch, professor and head of the Geography Department at the University of Wisconsin, Madison, and a co-worker of the *Atlas of World Agriculture*, wrote:

> . . . Dr. Baker was not much concerned over the popular but changing currents in geographical thought. He was concerned always about the many problems of regional economy, especially rural economy, and their inherent relation to the land. He sought in his studies to make clear by means of the written word and graphic device the nature and areal distribution of important associations of geographical facts. . . . To him no task seemed too great and no labor too hard when it was directed toward the interpretation of significant facts and associations in the area scene (Visher and Hu, 1950:331).

Unfortunately, Baker lacked the physical vigor to match his high mental ability and ambitious projects. His health was so poor during his youth that he did not start school until he was twelve. During the last decades of his life he usually appeared tired, but continued to work hard for long hours. His active mind envisioned numerous valuable studies, some of which he instigated or sponsored. He drove himself harder than his co-workers. He showed great appreciation for good work, and by so doing, encouraged others to do more and better work than they might otherwise have done. Throughout his professional life, Baker made a major contribution to geographic education in America.

Phillip Bacon

Phillip Bacon
July 10, 1922 -

Phillip Bacon was born and raised in Cleveland, Ohio. He began his college career at Citadel Military College in South Carolina in 1940. From 1942 to 1945 he was on active duty in the Navy, where he served as aerographer at several U.S. naval air stations. He received his B.A. degree in 1946 at the University of Miami, with a major in history and a minor in sociology, then went to George Peabody College for Teachers at Nashville, Tennessee, for graduate work. He was awarded an M.A. in history and geography in 1951 and an Ed.D. in social studies in 1955.

While in graduate school, Bacon started teaching social studies and history at the secondary school level. During the 1951 summer session he was instructor of history at George Peabody; during the summers of 1954 and 1955 he was instructor of geography. In 1955, after receiving his doctorate, he was appointed assistant professor of geography at the University of Pittsburgh. After two years in Pittsburgh he moved to New York City, where he was professor of geography and head of Graduate Studies in Geographic Education at Teachers College, Columbia University, from July 1957 to June 1963. At the same time he was professor of geography on the graduate faculty of Columbia University. He left Columbia University to serve as dean of the Graduate School at George Peabody College for Teachers during 1963-64. He then returned to Columbia University, where he held dual appointments as professor of geography in the Teachers College and in the Department of Geography at the University.

In 1966 Professor Bacon went to the University of Washington in Seattle, where he served as professor in the Department of Geography and professor of social studies in the College of Education (1969-71). In 1971 the University cited him for "distinguished undergraduate teaching." From 1967 to 1971 he was co-director (with John Jarolimeki) of the Tri-University Project in Elementary Education-Social Sciences (known as the TTT Project), which has had a continuing impact on the quality of elementary geographic education in the state of Washington. In 1971 he left Washington to become chairman of the Department of Geography and Anthropology at the University of Houston in Texas.

Bacon has been very active during his teaching career (James, 1974:62-63). He authored or coauthored twelve books and was editor or coeditor of twenty-nine others. His films and maps became important teaching materials. He was a member of the Steering Committee of the High School Geography Project, and served as a curriculum consultant for school systems in twenty-nine states and the District of

Thomas Frank Barton

Columbia.

His abilities, talents, skills, and accomplishments were recognized by the National Council for Geographic Education; he was elected second vice-president in 1964, first vice-president in 1965, and president in 1966. He served four terms on the NCGE's Executive Board.

Professor Bacon has received many honors and recognitions. He was awarded the National Council for Geographic Education Distinguished Service Award in 1974, and in 1975 and 1979 was awarded the University of Houston Teaching Excellence Award. In 1980 the University of Houston, College of Social Science, awarded him the Teaching Excellence Award. He was the first University of Houston faculty member to receive three awards for excellence in teaching. In 1976 he served as coordinator of the Regional Analysis team on the LACIE Project for NASA. In 1981-82 he served as president of the 27,000-member Alumni Association of George Peabody College/Vanderbilt University.

For two years Professor Bacon chaired the Publications Committee for the Association of American Geographers (1977-79). During 1976-79 he served on the AAG Council.

Thomas Frank Barton
December 3, 1905 - August 14, 1985

Thomas Frank Barton was born on the Barton Walnut Grove farm near Cornell, Illinois. The springs, streams, forest, hills, and wildlife on the farm provided an outdoor laboratory vastly more stimulating than the one-room rural school he attended. Upon graduating from the eighth grade, he was sent to live with his grandmother in the county seat of Pontiac while he attended high school.

In 1924 he enrolled at Illinois State Normal University, but dropped out of the university to teach two years (1925-27) in a one-room rural school. Later he returned to the university, where he received a B.Ed. degree in 1930.

Barton began teaching laboratory classes in geography and geology at Illinois State Normal University in the spring of 1930. In the fall he enrolled in the University of Wisconsin Graduate School as a geography teaching assistant, where he was assigned to assist Professor V.C. Finch, head of the Geography doing, encouraged others to do more and better work than they might otherwise have done. Throughout his professional life, Baker made a major contribution to geographic education in America.

In the summer of 1932 Barton taught two six-week sessions at Tennessee Teachers College (now Memphis State University). He was awarded a teaching assistantship in geography at the University of Nebraska in the fall of 1932 and enrolled in their Ph.D. program. Barton was invited back to Tennessee to teach a six-week session in the summer of 1933. He returned to the University of Nebraska as a teaching assistant in the fall.

Barton accepted a position during the spring of 1934 in the Social Studies Department of Nebraska State Teacher's College, Kearney. He returned to the University of Nebraska to continue writing his dissertation during the summer of 1934, and received his Ph.D. in January 1935. Barton was promoted from assistant to associate professor at Teacher's College, where he taught until September 1935. He then accepted a position as professor and head of the Department of Geography and Geology at Southern Illinois Normal University, Carbondale. After twelve years (1935-47) he accepted a position at Indiana University in September 1947, after serving as a visiting professor at the University of Nebraska during the summer.

At Southern Illinois University, Barton made geography a functional, living subject by recognizing the geographic problems of Southern Illinois and suggesting possible solutions, by publishing articles, and by grooming geography majors.

Soon after joining the Indiana University faculty, Barton was invited by the Department of State to be one of four delegates to the UNESCO Conference on the Teaching of Geography, held at McDonald College, St. Anne de Bellevue (near Montreal). He served on the International Geographical Union's Commission on the Teaching of Geography in Schools (1952-55) prior to going to Southeast Asia in June 1955.

A veritable traveler and internationalist, Barton was appointed professor of Social Sciences at the School of Education in Bangkok, Thailand, from June 1955 to June 1957. He made numerous field trips to all the countries in Southeast Asia except North Vietnam. With information gained from field studies, as well as library research, he wrote *An Economic Geography of Thailand: An Outline Study.*

Barton wrote twenty major articles on Southeast Asia and dozens of short articles about Thailand and Southeast Asia for books and journals. He also coauthored *Southeast Asia in Maps* (1970) and contributed articles to four other books about Southeast Asia — *Readings in Economic Geography* (1967), *Methods of Geographic Instruction* (1968), *World Geography* (1972), and *Southeast Asia, Realm of Contrasts* (1974).

During his three consecutive five-year elected terms as editor of the *Journal of Geography* (1950-65), Barton made it an international periodical. In 1965 he was invited to join the Southeast Asia Development Advisory Group (SEDAAG), which was sponsored by the Asia Society and the Agency for International Development.

At Southern Illinois, Barton supervised the establishment and management of a U.S. Airway Weather Station in the Geography Department. As a consultant for Coronet Film Corporation, he collaborated on the production of eleven educational films. He became a consultant for Indiana University's Audio-Visual Center, and served as consultant for Universal Education and Visual Arts (1966-77). His articles about various visual aids appeared in *Education, Journal of Geography, Social Education,* and other magazines.

Even though Barton was always deeply engrossed in teaching, he found time to write and edit.

> A versatile writer, he has illuminated a wide range of topics for teachers at all levels from elementary to college. . . . He was among the first, if not the first, to design and write texts in geography for the primary grades (Cunningham, 1965:423).

Barton contributed two new concepts to urban geography — "The Commercial-Residential Ecoton," *The Scientific Monthly,* 1937, Volume 44, pages 131-136; and "The Urban Web Hypothesis," *Interplay: The Magazine of International Affairs,* 1970, Volume 3, pages 43-47. Senator Hubert H. Humphrey wrote the introduction for the latter, and it was reprinted in the *Congressional Record,* October 8 1971, pages 35, 619-635.

Barton served the National Council for Geographic Education with distinction. He served as president of three state councils — Nebraska (1935), Illinois (1937-38), and Indiana (1962). He was elected secretary of the National Council of Geography Teachers in 1940 and served a five-year term; he was elected second vice-president in 1946, first vice-president in 1947, and president in 1948. He became assistant editor of the *Journal of Geography* in 1948, and in 1950 was elected to the first of three five-year terms as editor. He edited the *Journal of Geography* longer than anyone except George J. Miller.

Barton was active in Gamma Theta Upsilon. He was inducted into

Charles Carlyle Colby

the Alpha Chapter at Illinois State University. He established the Lambda Chapter at Southern Illinois University and the Tau Chapter at Indiana University. He was elected and served two years (1946-50) as national first vice-president in charge of establishing new chapters.

In 1965 Barton received the National Council for Geographic Education Distinguished Service Award. The Alumni Association of Illinois State University awarded him the Distinguished Alumni Award in Education in 1975. In 1976 he received the Rocking Chair Award from Sigma Delta Chi of Indiana University for major contributions to Indiana University, and the same year was named professor emeritus. In 1977 he received an honorary Doctor of Law degree from Illinois State University, and in 1978 he was awarded a Distinguished Service Award by the Geographical Society of Chicago (Black, 1986:42-43).

For over forty years, Barton enthralled students with his thought-provoking lectures. This writer (Karan) had the privilege of attending Barton's stimulating lectures on environmental conservation and urban geography as a graduate student at Indiana University in 1954. Twenty-two years later, in 1976, the year before his retirement, hundreds of students sought to enroll in his courses at Indiana University.

Charles Carlyle Colby
April 13, 1884 - July 16, 1965

As a student and faculty member, Charles C. Colby understood the importance of excellence in teaching. Born in Romeo, Michigan, his academic career began at Michigan State Normal College under Mark Jefferson, one of the most stimulating and creative teachers of geography in the United States. After earning his bachelors degree at Michigan State Normal College in 1908, Colby continued his academic career at the University of Chicago, where great figures in the field at that time — Rollin D. Salisbury, J. Paul Goode, Harlan H. Barrows, Ellen Churchill Semple and Walter S. Tower — were teaching.

After receiving a B.S. degree from the University of Chicago, Colby became an instructor and head of the Department of Geography at Minnesota State Normal School in Winona, where he became interested in the historical geography of southeastern Minnesota. His first paper on this subject was presented at the 1912 meeting of the Association of American Geographers at Yale University.

At the time, George Peabody College was attempting to raise teaching standards in the southern United States. In 1914 Colby was invited to develop a geography program at George Peabody.

After hearing Colby's paper on the historical geography of southeast Minnesota, Professor Salisbury of the University of Chicago suggested this topic as a suitable subject for a doctoral dissertation. In 1916 Colby returned to the University of Chicago, completed his doctorate, and remained as a member of the faculty for the next thirty-two years.

Professor Colby was active in the early years of the National Council of Geography Teachers. He is listed among the nine elected members of the Executive Committee in the *Journal of Geography,* October 1916, and in 1924 he was elected to the Executive Committee for a three-year term. "Colby maintained an interest in the teaching of geography at all levels at a time when this was not fashionable in graduate schools" (Harper, 1966:296-297). In 1931, with Alice Foster, Colby wrote a standard economic geography which was used in high schools for more than two decades.

A contemporary has written about Colby:

> He had an abiding concern for students and the rare ability to discover the strong points of a person or a term paper. Thus his students regularly accomplished more than even they felt capable of doing (Harper, 1966:296).

In 1952 Colby received the National Council for Geography Teachers Distinguished Service Award.

Colby's career may be divided into four parts (Harris, 1966:378-382). First, as a student and faculty member. Second, from 1921 to 1935, a period when he wrote scholarly articles and two books. This period culminated in his presidential address before the Association of American Geographers in 1935. Third, a period when he wrote papers and monographs resulting from government-sponsored research projects, such as the Land Committee of the National Resources Planning Board. This fourteen-year period ended with his retirement from the University of Chicago in 1949. The fourth period, which lasted sixteen years, was spent as a visiting professor at various colleges and universities such as the University of Illinois at Urabana, the University of California at Los Angeles, Michigan State University, and Southern Illinois University. At Southern Illinois University he published *Pilot Study of Southern Illinois* and also directed the Mississippi Valley Investigations.

Many honors, recognitions, and responsibilites were awarded to Colby. In addition to being president (1935) of the Association of American Geographers, he was secretary from 1921 to 1928. He was president of the Geographic Society of Chicago, and received the Society's Helen Culver Gold Medal. He also held honorary degrees from both Michigan State Normal College and Southern Illinois University.

On the evening of July 15 1965, at the age of eighty-one, Colby called his family from La Crosse, Wisconsin, to report on how well his field work was going. The next morning he did not wake up.

George Babcock Cressey

George Babcock Cressey
December 15, 1896 - October 21, 1963

George Cressey possessed a superior mind, enormous energy, and an insatiable desire to learn geography by observation. He was destined to become one of the world's leading authorities on the land mass of Eurasia. *The New York Times* wrote:

> "No other living geographer has seen so much of the great continent of Asia as Dr. Cressey."

Professor Cressey's father was Dr. Frank G. Cressey, a Baptist minister who was associated with President William Raney Harper at the opening of the University of Chicago in 1892. George's mother, who was the first woman to graduate from the University of Chicago, taught Latin at Dennison University, where George received a B.S. degree in 1919 with a major in geology. He went to the University of Chicago for graduate work and received his Ph.D. in geology in 1923.

The same year Cressey was recruited by the American Baptist Mission Society to teach at Shanghai College. Instead of traveling by boat, he went overland across Asia and entered China after crossing Siberia, the Gobi, and Mongolia.

During his first year in China, Cressey attended the College of Chinese Studies in Peking, where he found that a young American student, Marion Holbrook, had an adjacent mailbox. She was also studying Chinese, preparing to become a social worker in Tientsin for the Congregational Mission Board (James, 1964:254-257). In 1925 they were married.

As a professor in Shanghai, Cressey had the opportunity to study China in the field. Between 1923-29 he traveled over 30,000 miles on reconnaissance trips into twenty-eight provinces and all of the country's geographic regions. As he traveled he made copious notes, took photographs, and both made and collected maps. After using these materials in his teaching, he arranged them in book form. When he left China in 1929, his completed manuscript, *The Geography of China,* was left with the Commercial Press of Shanghai for printing.

However, wars kept interfering with Cressey's field research and publications. His reconnaissance study was frequently interrupted by encounters with bandits and the crossfire of armies engaged in civil wars. In 1932, during the Japanese invasion of China, the Commercial Press and Cressey's manuscript were destroyed by a fire, and with it nearly a decade of his work.

Fortunately, he still had his field notes. With these, and partly from memory, he rewrote the manuscript. In 1934 *China's Geographic Foundations* was published by McGraw-Hill Book Company.

As he studied in the field, observed the landscape, and concentrated on the problems of man's use of the land and his habitat, Cressey's interest changed from geology to geography. In 1929 he returned to the U. S. to study geography at Harvard and Clark universities. In 1931 he received a second Ph.D. in geography from Clark University.

In September 1931, at the age of thirty-five, Cressey was appointed head of the new Department of Geology and Geography at Syracuse University. In addition to teaching undergraduates, he developed at Syracuse what was widely recognized during the 1930s as one of the best Master of Arts programs in geography in the country.

In 1937 Cressey's insatiable desire to learn by traveling took him to Moscow to attend the International Geological Congress. He was invited to Moscow as a consultant on the Great Soviet World Atlas. He traveled for three and a half months in Siberia and in the Arctic. His second trip into Siberia allowed him to observe its tremendous economic growth and potential.

A charming and fluent speaker, Professor Cressey was constantly asked to lecture on the Soviet Union and China. His remarks and opinions were substantiated with statistics and visuals such as graphs, maps, and photographs. International geographers respected him for his honest conclusions based on field work and scholarship, but some political personalities in the U.S. mistook his reporting as praise for the communist regimes in the USSR and China. His name appeared from time to time on the list of scholars suspected of sympathizing with the Communists. His name also appeared on a list of enemies published by the Chinese Communists, who included Cressey's name among the most dangerous tools of Wall Street and the warmongers.

> America's involvement in World War II brought Cressey to Washington as a consultant to the Department of State, the Board of Economic Warfare and Army Intelligence, and to Chungking in 1943-1944 for the National Academy of Sciences in order to help transplanted Chinese universities (Herman, 1965:360-364).

During the War Cressey was kept busy teaching in the accelerated Army training programs at Syracuse and lecturing on the Far East. During this strenous period he completed his second book, *Asia's Lands and People.*

In 1945 Cressey became the first chairman of the new Department of Geography at Syracuse. His special goal was to bring Asian geographers and graduate students to Syracuse. He was active in broadening the work of the Association of American Geographers, the International Geographical Union, and the Association of Asian studies. The world-wide extent of these organizations seems natural today, but this is due to a few pioneers like Cressey who realized that scholars are found throughout the world and that support for scholarly international research and reporting can only come when the general public has a similar outlook.

In 1951 Cressey's retirement from the chairmanship at Syracuse at the age of fifty-five and his election as a Maxwell Professor gave him time during the following decade for extensive travel, field study, writing, and lecturing. He focused his attention on Southwest Asia, and his widespread field studies in that part of Asia are embodied in his book, *Crossroads: Land and Life in Southwest Asia,* published in 1960. *Soviet Potentials* was also published that year.

Professor Cressey's continued enthusiasm for travel and field study, coupled with vigorous physical energy, resulted in the publication of ten books and nearly 200 articles. A bibliography of his publications appears in the *Annals,* June 1965, pages 362-364. "Some of his books have been through numerous editions and have been translated into many languages" (James, 1964:256).

Cressey was elected to Phi Beta Kappa and Sigma Xi. He was president of the International Geographical Union (1949-52) and served the Executive Committee of the International Council of Scientific Unions (1949-52). In 1952 he received the George Davidson Medal from the American Geographical Society. In 1957 he was honorary president of the Association of American Geographers, and the same year was awarded the National Council for Geography Teachers Distinguished Service Award. He served as a Fulbright research scholar in Iraq in 1955-56 and as a professor in Lebanon in 1957-58. From 1952 until his death he was a member of the Council of the American Geographical Society. He was president of the Association for Asian Studies (1959-60) and was appointed its National Visiting Scholar in 1961-62. In 1948 he was awarded the honorary Doctor of Human Letters degree by Dennison University.

Alden Cutshall

Alden Cutshall
April 12, 1911 -

Alden Cutshall was born in Olney, Illinois, thirty miles west of Vincennes, Indiana. Although he has lived most of the fifty-two years of his professional life in Illinois and the Midwest, his activities and publications substantiate that he is an internationalist in both mind and spirit.

Cutshall's university education started at Eastern Illinois University, where he received a B.Ed. degree in 1932. He began his teaching career as a teacher in Richland County, Illinois, teaching five of the eight elementary grades. He received his M.A. in 1935 from the University of Illinois and his Ph.D. from Ohio University in 1940. He then began forty years of service at the University of Illinois. After teaching in Urbana (1940-47), he transferred to the Chicago Undergraduate Division of the University of Illinois (1947-65), then taught for sixteen years at the Circle Campus (1965-80). He served as head of the Department of Geography during 1964-69. In 1972 he was appointed professor emeritus, with the provision that he retain an office and teach at least one class each year on a part-time basis, an arrangement which continued until 1980.

Because of his superior teaching and other talents, Professor Cutshall was invited to teach at other universities. After teaching during the 1941 summer session at Ball State Teachers College, Muncie, Indiana, he taught at George Peabody College for Teachers, Nashville, Tennessee, during the summer of 1943. He taught at the International College in Chicago (1948-49); at the Institute of Asian Studies, University of the Philippines (1957-58); at Aurora College in Illinois (1973-74); and at Northern Illinois University (1977).

Cutshall's other professional experiences include working for federal, state, and international organizations. He served in the Far Eastern Division, Office of Strategic Services, Washington, D.C. (1944-45), and in the Department of State (1945-46). During May-June 1945 he was attached to the Military Government Group, Monterey, California (detached duty from OSS), and served in Manila during the summer of 1951. Professor Cutshall held geography workshops for the Illinois Office of the State Superintendent of Public Instruction in 1966, and for Western Illinois University during the summers of 1966 and 1967.

Although Cutshall's publications and teaching covered a broad range of geographic topics, he showed a special interest in the geography of the Philippines and published his first professional article, "Trends in the Philippine Sugar Industry" in *Economic Geography* in 1938. His

Elizabeth Eiselen

teaching, publications, and continued interest in this part of the world earned him a Fulbright research grant in 1950, with the opportunity to serve in Manila during the summer of 1951 and again as a Fulbright lecturer in 1957.

In 1964 he published his first book, *The Philippines: A Nation of Islands.* During 1963-65 he was associate editor of the *Philippine Geographical Journal.* He continues to promote geographic understanding of that island country and has contributed chapters on the Philippines to the textbooks, *World Political Geography* and *Southeast Asia: Resources and Potentials.*

Since his retirement, Cutshall has conducted research and has revised the Philippine chapter in *Southeast Asia: Realm of Contrasts* and the historical geography chapter in *Illinois: Land and Life in the Prairie State.*

During his long professional career, Professor Cutshall has received many grants and honors. In addition to the Fulbright research grant to the Philippines, the Social Science Research Council awarded him a travel grant to the International Geographic Congress, Stockholm (1960); and the Society of Chicago awarded him a travel grant to the International Geographic Congress in Moscow for the 1976 meeting.

In 1983 the National Council for Geographic Education awarded him the George J. Miller Distinguished Service Award.

Elizabeth Eiselen
August 22, 1910 - January 27, 1990

Elizabeth Eiselen was born and raised in Evanston, Illinois. She became interested in geography as an undergraduate at Northwestern University, where she earned her B.A. in 1931. She studied under J. Russell Smith at Columbia University and earned her M.A. degree there. She transferred to the University of Chicago and studied under Harlan H. Barrows, Charles C. Colby, Henry M. Leppard, Edith P. Parker, and Robert S. Platt. She received a Ph.D. in 1943.

Eiselen's teaching career began at Union College in mountainous southeastern Kentucky (1934-37). She then went to Wellesley College, Wellesley, Massachusetts, as an assistant professor in 1942, and moved up through the ranks to become professor and chairman. After returning to Chícago in 1965, she taught at Concordia Teachers College, River Forest, and at De Paul University (Houk, 1990:113).

Sidney Ekblaw

To enrich her understanding of the environment and peoples of the earth, she traveled to all parts of the world. She then shared these experiences in the classroom and more widely in articles and textbooks.

Eiselen's publications range from articles in professional journals to topics in encyclopedias and yearbooks to textbooks. She was the senior author of junior high school textbooks on Africa and Latin America. Eiselen prepared a series of six educational filmstrips on the geography of Massachusetts, and served as a consultant on geography programs.

She shared her talents, skills, and energy in geographical and public organizations. She served as New England representative on the Council of the Association of American Geographers; as a member of the Executive Board of the Society of Women Geographers; as president of the New England-St. Lawrence Valley Geographical Society (twice); and as first vice-president of the Geographical Society of Chicago. She was also an active member of the Illinois Geographical Society and the De Paul Geographical Society.

Professor Eiselen is known among the members of the National Council for Geographic Education for her twenty years (1958-78) of service (*Journal of Geography,* 1978:248). She served almost continuously from 1957 on the Executive Board. She chaired and served on major committees, contributed to the editorial work of the *Journal of Geography,* especially when its editorial office was located at Concordia Teachers College, and was manuscript editor of NCGE publications. She served as treasurer from 1961 to 1966, relinquishing the position to become the first executive secretary of the NCGE. She later served as second vice-president, president-elect, and president in 1975. In 1978, near the conclusion of her post-presidential term on the Executive Board, she received the National Council for Geographic Education George J. Miller Award for Distinguished Service.

Sidney E. Ekblaw
March 24, 1903 -

Sidney E. Ekblaw — geographer, author, professor and chairman of his department at the University of Missouri-Kansas City — has exerted leadership in geographic education for nearly fifty years. Through his inspirational teaching and conscientious efforts on behalf of the continued development of excellence in teaching, he has demonstrated superior service in the field of geography.

Sidney grew up on an Illinois farm on the rolling terrain of a recessional moraine, four miles northeast of Rantoul, Illinois. He attended a one-room country school through the eighth grade, and graduated from Rantoul Township High School. He received his A.B. degree in 1929 and M.S. degree in 1930 from the University of Illinois at Urbana. He studied at Clark University for his Ph.D. in geography, where he was also editorial assistant for *Economic Geography* during 1933-34.

Joining the University of Kansas City faculty in 1934, Ekblaw initiated a variety of activities to advance geographic education. He helped organize the Missouri Council for Geographic Education, promoted state meetings, and circulated newsletters. Through monthly meetings and field trips of the Heart-of-America Geology Club, he interested numerous persons in the earth sciences.

During his thirty-six year association with the University of Kansas City (which became the University of Missouri-Kansas City in 1963), Ekblaw promoted geography across the state, developing programs and courses of instruction. He was the guiding hand in the formation of the Department of Geology-Geography, now called the Department of Geosciences. As professor of geography and geology, he touched the lives of thousands of students, many of whom now hold important positions in various fields.

Off campus, Ekblaw promoted geographic education through lectures to elementary school classes, high school assemblies, science clubs, and service organizations. He helped organize and promote the Kansas City seminar of the American Meteorological Society in 1934-35, the first of many such chapters now in the U.S. He led numerous geographic field trips, including a tour of South America during the International Geographical Union meetings in Rio de Janeiro in 1956. He concluded his teaching career with three years of part-time instruction at Arizona State University from 1970 to 1973.

In retirement in Scottsdale, Arizona, Sidney Ekblaw continues to take an active interest in geography and geographic education. He regularly attends meetings of the National Council for Geographic Education, and has long been one of its strongest supporters. He served the Council as second vice-president in 1961, first vice-president in 1962, and president in 1963. In an even greater role, though often not noticed, he served the Council in support and leadership whether or not he was actually in office.

Apart from his service to the National Council for Geographic Education, his influence has also been felt through articles in professional journals, including *The Journal of Geography,* and in high school and college textbooks. Whether writing about grasslands, Carlinville limestone, conservation of fuel resources, or social and economic geography, he is always able to explain complex processes in readily understood terms. Moreover, his style displays a philosophic bent, an acuity of perception, and a spark of humor that add up to delightful reading. He is one of the few geographers with a gift for clear and literate exposition.

The National Council for Geographic Education conferred the George J. Miller Distinguished Service Award on Sidney E. Ekblaw in 1982.

Edward B. Espenshade

Edward Bowman Espenshade
October 23, 1910 -

Edward Espenshade was born in Chicago. Except for World War II service in Washington, D.C., he never left his state for very long. He was awarded his B.S. and M.S. in geology and his Ph.D. in geography from the University of Chicago. He began his professional career as a junior geologist with the Illinois Geological Survey in 1932, and spent the next eleven years at the University of Chicago, first as a curator and later as an instructor.

Following wartime service as foreign map editor for the Army Map Service and as an intelligence specialist for the U.S. Army Corps of Engineers, he joined the Northwestern University faculty. He taught a variety of courses, including a physical geography class that usually attracted more than 300 students. He taught at Northwestern for thirty-five years, becoming a full professor in 1955 and serving as departmental chairman between 1958 and 1977.

> Despite his close association with other traditions in geography, Professor Espenshade had the tolerance, the vision, and the innovative boldness to open the doors of his department to the new wave of quantifiers and theoreticians (*Journal of Geography*, 1979:248).

As a result of Espenshade's skillful administration, Northwestern emerged as a major center of the "new geography" without losing its competence in traditional geography.

Although Professor Espenshade's special interests included physical geography and Asia, he received national acclaim in cartography. His name and *Goode's World Atlas* have become virtually synonymous. He contributed numerous maps to the Atlas and later became its chief editor. Under his imaginative and innovative guidance, *Goode's World Atlas* has become the most widely used classroom reference of its kind in the country. Espenshade also edited — with the cooperation of associate editors — the Japanese, Italian, Spanish, and British versions of *Goode's School Atlas.*

Espenshade received many honors and recognitions during his long professional career in geography. In 1942 the University of Chicago conferred upon him an undergraduate teaching award. In 1964 he served as president of the Association of American Geographers, and he was president of the Chicago Geographical Society during 1969-71. He also served as chairman of the Earth Sciences Division of the National Research Council-National Academy of Sciences, the AAG Commission on College Geography, and on numerous other professional

Otis W. Freeman

committees. He was appointed professor emeritus at Northwestern University on January 1, 1979. In recognition of his many contributions to geography and his activities in promoting geography, the National Council for Geographic Education awarded him the George J. Miller Award for Distinguished Service in 1979.

Otis Willard Freeman
April 13, 1889 - April 3, 1964

Teaching, research, publications, administration, and the energetic promotion of geography characterized the career of Otis Freeman.

Freeman was born in Ostego, Michigan. He graduated from Albion College in 1910 and completed his M.S. in geology at the University of Michigan in 1913. During the next twelve years he taught physical science at high schools in Michigan, Montana, and California. In 1924 he joined the faculty of Eastern Washington State College (then known as Cheney Normal) and concentrated on teaching geology and geography. In 1929 he completed his Ph.D. at Clark University.

At Eastern Washington College of Education, Freeman became professor of geography, head of the Science Department, and eventually, president of the college in 1951. After retiring from administrative work, he returned to teaching, spending four years at State Teachers College in Indiana, Pennsylvania (now Indiana University of Pennsylvania), followed by a year at the University of Arizona. During his thirty-three year career he held visiting professorships and/or summer posts in Indiana, Hawaii, San Francisco State, and Washington (Martin, 1964:332).

Freeman's published research on the Columbia Basin was followed by regional and economic studies of the Pacific Northwest and later of Hawaii and Alaska. His books include *Economic Geography of Hawaii, The Pacific Northwest* (with Howard H. Martin), *Essentials of Geography* (with Hallock F. Raup), *Resources of Washington State* (with Roland Upton), and *World Geography* (with John W. Morris). Freeman's bibliography includes some 125 contributions which appeared in eleven journals. In addition to geographical publications, he contributed to various geology, meteorology, mining, and engineering journals. At various times he served as associate editor of the *Journal of Geography, Economic Geography,* and the *Pacific Northwest Quarterly.* He was geography editor of *Education* for eight years.

After holding various offices in the National Council for Geographic Education, Freeman was its president in 1953-54. Four years later, at the meeting of the Council in Buffalo, New York, he was awarded the National Council for Geographic Education Distinguished Service Award. In 1957, when he retired at Eastern Washington, he was designated professor emeritus of geography and president emeritus of the college.

Freeman played a major role in founding the Association of Pacific Coast Geographers in June 1935, becoming its first president and publishing the first *Yearbook,* of which he was editor for five years. The APCG designated its 1964 Berkeley meeting as a memorial to Freeman.

A colleague of Professor Freeman, Howard H. Martin of the University of Washington, describes Freeman thus:

> Kindly and constructive in all of his relationships, Otis Freeman left a lasting imprint on successive generations of students and scores of his colleagues (Martin, 1964:332).

Erna Grassmuck Gilland

Erna Grassmuck Gilland was born in Philadelphia and spent her entire professional life in Pennsylvania. She not only taught geography but engaged in numerous administrative activities which, along with the efforts of other leaders in geographic education, resulted in Pennsylvania becoming one of the leading states in the development of geography in its elementary schools, high schools, and universities.

> Gilland believed that there was no substitute for good geography teaching. Gilland constantly worked for more effective methodology while insisting that method is not enough; that without rich and stimulating subject matter, there is no geography (Freeman, 1954:364).

During Gilland's busy professional career she served at various times as a teacher, in-service supervisor, professor, and state director of geography. She helped train teachers, organized geography clubs, and addressed countless institutes and meetings.

In the 1920s she frequently attended the annual meetings of the National Council for Geography Teachers. She was more prone to express her ideas orally than in publications. She is listed as one of the coauthors of Chapter XV, "Geography in the Elementary School," in the 32nd Yearbook, *The Teaching of Geography*, by the National Society for the Study of Education, but the Index to the *Journal of Geography*, Volumes XXI-LV (1922-56) indicates that she contributed only one review to the official journal of the NCGT. She was an associate editor of the *Journal of Geography* in the 1930s, and was elected second vice president of the National Council for Geography Teachers in 1923, first vice-president in 1925, and president in 1926.

Gilland served as secretary of the geography section of the World Federation of Education (later the World Organization of the Teaching Profession), received national recognition as an authority on geography curriculums, and was a highly-respected consultant on geographic education for over a third of a century.

In 1954 a committee of her peers selected Gilland to be the recipient of the National Council of Geography Teachers Distinguished Service Award, citing her dynamic qualities. They stressed that when she began teaching, geography in the public schools was traditionally dull and factual.

> With her characteristic drive and enthusiasm she began training teachers and stirring up superintendents. Soon, Pennsylvania was one of the states leading in geography-consciousness. This influence gradually spread to other areas and her name became synonomous with better geography teaching in America. She proved that geography can be the most stimulating subject in the curriculum, a real link with life. Some measure of her enthusiam has been imparted to the many geographers with whom she worked (Freeman, 1954:365).

Herbert Henry Gross

Herbert Henry Gross
March 14, 1905 - September 10, 1981

Herbert Gross was born and raised in Buckley, a village in Eastern Illinois about 100 miles south of Chicago. His undergraduate education was acquired at Concordia Teachers College and the University of Chicago. He earned his degrees — the Ph.B. (1929), M.A. (1934), and Ph.D. (1948) — at Chicago. He also did graduate work at the University of Illinois, Northwestern University, and at Lewis Institute in Chicago.

His teaching and administrative career began in 1925 in St. John's Lutheran School in Buckley. Upon receipt of his bachelors degree in 1929, he became associated with St. Luke's Lutheran School in Chicago, where he served for eleven years as both teacher and principal. From 1937 to 1940 he was employed as a part-time instructor at Concordia Teachers College while still maintaining his elementary school connections (Phelps, 1970:558).

Professor Gross became permanently associated with Concordia Teachers College as a professor of geography in 1940, and in September 1969 was appointed head of the Department of Geography. He also taught summer school at the University of Chicago and at the University of Washington. He participated in three NDEA Geography Institutes; in two of them he was a teaching faculty member, in the third he was the director.

The contributions of Herbert Gross go beyond instruction and administration. He also served as an evaluatior of educational research projects and as a consultant to the National Science Foundation.

As a writer he authored, coauthored, edited, and co-edited a score of books and monographs. In addition, he served eleven years as the editor of *Lutheran Education* and five and one-half years as editor of the *Journal of Geography*.

For over two decades (1950-70), Professor Gross made major contributions to the growth of the National Council for Geographic Education. His first article, "Indices of Good Geographic Instruction," was published in the *Journal of Geography* in 1952. He served as the Illinois coordinator of the NCGE (1955-61), as second vice-president (1962), first vice-president (1963), and president in 1964. In 1965 he began a five-year term as editor of the *Journal of Geography* (Gabler, 1982:78-79).

In June 1968 Valparaiso University granted Professor Gross a Doctor of Law degree, and in October 1968 Bethany College of Lindsburg, Kansas, awarded him the Doctor of Letters degree.

Richard Hartshorne

Richard Hartshorne
December 12, 1899

Richard Hartshorne was born in Kittanning in west central Pennsylvania, thirty-five miles northeast of Pittsburgh. He was a student at Yeates School, Lancaster, Pennsylvania (1912-16). He received a B.S. from Princeton University in 1920 and a Ph.D. from the University of Chicago in 1924. Clark University granted him an Ll.D. in 1971.

Hartshorne was an instructor of geography at the University of Michigan (1924-27), an assistant professor (1927-37), and an associate professor (1937-40). In 1940 he joined the faculty of the University of Wisconsin as an associate professor, became professor in 1941, and professor emeritus in 1970. He served as chairman of the Department of Geography from 1950 to 1954. He was on the faculty of Pennsylvania State University as a visiting professor in 1965.

In 1918, during World War I, Hartshorne served as a private. During World War II, while on leave (1941-45), he worked with the Office of Strategic Services (O.S.S.) in Washington as chief of the Geography Division, then became assistant chief (research director) of the Research and Analysis branch. He also served on the civilian faculty of the National War College.

Before the start of World War II, Hartshorne was a fellow of the Social Science Research Council (1931-32). He was a member of the Council of the Association of American Geographers during 1937-40, and was president in 1949. He was vice-chairman of the geography section of the Division of Geography and Geology of the National Research Council during 1940-42, and was chairman of a section on methodology and bibliography of the International Geographical Congress held in Stockholm in 1960.

Professor Hartshorne received national and international recognition with his books, *The Nature of Geography* (1939) and *Perspective on the Nature of Geography* (1959). One of his more recent books is *The Academic Citizen* (1970). Hartshorne had a major impact on geographic education through his numerous publications, particularly *The Nature of Geography*, which is one of the most influential works in American Geography (Entrikin and Brunn, 1989).

In 1960 Hartshorne was the recipient of the Daly Gold Medal from the American Geographical Society and the Achievement Award of the Association of American Geographers.

Preston E. James

Preston Everett James
February 14, 1899 - January 5, 1986

Preston E. James was born, raised, and received his pre-college training in Brookline, Massachusetts. He was awarded an A.B. in 1920 and an A.M. in 1921, both from Harvard University, and a Ph.D. in 1923 from Clark University. He started his university career as a teaching assistant in geography at Harvard University (1919-21). He taught at Radcliffe College in 1920, was an instructor at Clark University (1921-23), then went to the University of Michigan as an instructor (1923-24).

James rose through the ranks at Michigan and became professor of geography in 1934, holding this rank until 1945, when he joined Syracuse University as professor of geography. At Syracuse he was chairman of the Geography Department from 1951 to 1968. In 1961 he was appointed the Frank Smalley Professor of Geography, a position he held until 1964, when he was appointed the Maxwell professor of Geography. In 1970 he became professor emeritus.

Professor James served as a geographic consultant for the Comselho Nacional de Geografia, Brazil (1949-50); chief, Latin America Division, Office of Strategic Services (1941-43); and chief, Europe-Africa Division, Office of Strategic Services, (1943-45). He was a member of the National Research Council (1937-40), vice-chairman, Division of Geology and Geography (1939-40); and member of the Commission on Geography of the Pan-American Institute of Geography and History (1949-57).

During a prolific lifetime, Professor James received numerous recognitions and awards. He served as secretary of the Association of American Geographers (1936-41) and as president in 1951. He was a member of the Council for Latin American Affairs and became president in 1957. He received a grant-in-aid for field research in Latin America from the National Research Council in 1930 and from the Social Science Research Council in 1938. He served as summer lecturer at the University of California (1933), the University of Wisconsin at Madison (1937), the University of Texas (1940), Harvard (1941), and at the Institute of Advanced International Studies (1946).

James received the National Council for Geographic Education Distinguished Service Award in 1964. The Pan American Institute Medal of Geography and History was awarded to him in 1965, and the following year he received the Livingston Centenary Medal from the American Geographical Society. The Helen Culver Medal was awarded to him by the Geographic Society of Chicago (1973) and the Latin America Geographic Award was conferred in 1977. In 1979 he received the

Clyde F. Kohn

National Council for Geographic Education Master Teacher Award in Mexico City and in 1980 was awarded the Royal Geographical Society Medal in London.

James held four honorary degrees. The Doctor of Science degree was conferred by Eastern Michigan University (1967); the Doctor of Laws was issued by Clark University (1968); the Doctor of Human Letters was conferred by Syracuse University (1973); and the University of Michigan awarded him a Doctor of Laws degree (1974).

After retirement, James coauthored *The Association of American Geographers* with G.J. Martin in 1978. With authors Kempton Webb and Eileen James, he coauthored *One World Divided* (1980). The following year *Global Geography,* with coauthor Zelda Davis, was published, and *All Possible Worlds* was revised with coauthor G.J. Martin.

In replying to Barton's letter of inquiry, James wrote:

> Unfortunately, I have been declared legally blind because of a disease called muscular degeneration so that reading is very difficult and writing is impossible. My wife Eileen handles all my mail and also helps me with revisions. . . . (letter from James to Barton, August 9 1982).

James's last article, "The University of Michigan Field Station at Mill Springs, Kentucky, and Field Studies in American Geography," appeared in *The Evolution of Geographic Thought in America: A Kentucky Root,* edited by Wilford A. Bladen and P.P. Karan, Kendall-Hunt Publishing Company, Dubuque, Iowa, 1983.

Clyde Frederick Kohn
April 10, 1911 - November 22, 1989

Clyde Kohn was born and raised in Mohawk, Michigan, a village located ten miles northeast of Houghton in Michigan's upper peninsula. After five years of summer school interspersed between teaching in elementary and secondary schools in Michigan, he earned the B.A. degree in 1935 from Northern Michigan University after one year of residence. With a fellowship for graduate study at the University of Michigan, he had to decide whether he should major in elementary education or in geography.

Due in part to the example and advice of two enthusiastic members of the National Council of Geography Teachers (now the National Council for Geographic Education), J. Russell Whitaker and Preston E. James (Houk, 1973:61-63), Kohn decided to specialize in geography, and received the M.A.in 1936 and Ph.D. in 1940. He started his teaching career at Mississippi State College for Women, then moved to Harvard University, where he spent three years (1942-45) working under Derwent C. Whittlesey.

In 1945 Professor Kohn joined the Department of Geography at Northwestern University, where his expertise and experience in elementary education resulted in his appointment as associate professor of geography and education (1948-53). Thanks to that arrangement, he was director of the School of Education's programs for teaching social studies and geography in the elementary schools until 1958, when a change of policy discouraged Northwestern's liberal arts faculty from taking an active interest in the educational problems and teaching methods of the elementary schools.

Professor Kohn next accepted a position at the University of Iowa, where his interest in geographic education was encouraged by Harold McCarty. Kohn became chairman in 1965 and began a program of expanded departmental offerings which stressed quantitative techniques and theoretical geography. He retired as chairman in December 1978.

Involvement in improving geographic education goes far beyond classroom contributions. Professor Kohn served as co-chairman of the NCGE-AAG committee which initiated the High School Geography Project (1960-63). Later he served as chairman of the Office of Education's panel, Project Impact, which was set up to study the impact of NDEA institutes on teacher education. He served as geographic education consultant to various school systems and to several of the nation's leading map-making and audio visual manufacturers. He also authored and/or edited twenty-four textbooks for primary and secondary schools.

Professor Kohn's research and writings were not limited to geographic education. Soon after joining the Northwestern University faculty, he served as director (1949-55) of an Office of Naval Research project, which experimented with different types of keys for the identification of land use patterns on aerial photographs. While director of the Center for Metropolitan Studies at Northwestern (1954-58), he was involved in urban geography and co-edited *Readings in Urban Geography*. He wrote fifty-six articles and chapters in books on population distribution and settlement geography.

Professor Kohn worked diligently for the National Council for Geographic Education for four decades. After serving as secretary for three years (1946-48), he served as acting president in 1951 following President Harry O. Lathrop's death, and was elected president in 1952.

He was a member of the Nominating Committee (1977) and chaired the Mission Committee (1978). He served as consulting editor of the Pacesetters Series (1977-83). In 1981, at the annual meeting of the NCGE in Pittsburgh, he presented a paper, "Looking Back and Looking Ahead," which was later published in the *Journal of Geography*. He served on many other NCGE committees and projects, was state coordinator for Iowa, and was also a member of the Editorial Board of the *Journal*.

Kohn also budgeted his time so he could be active in many professional organizations, especially the Association of American Geographers, of which he was vice-president (1966-67) and president (1967-68). He had the distinction of being president of both the NCGE and AAG. Many AAG committees were chaired by him, including the Task Force on Graduate Education and the Commission on Geographic Education. He also served the National Council for Social Studies as a member of its executive and editorial boards and was editor of the 1949 Yearbook. The National Council for the Accreditation of Teacher Education listed him as a member of key committees.

Kohn was active in international organizations concerned with geographic education. He was the U.S. correspondent for the International Geographical Union's Commission on Teaching Geography, and was the Association of American Geographers' representative to UNESCO (1961-66).

Professor Kohn, upon leaving the University of Iowa faculty in December 1979, was appointed professor emeritus in 1980. He was visiting professor at the Department of Geography, University of Hawaii, the spring of 1980. He also served as a member of the Planning Committee of Unified Science and Mathematics in the Elementary Schools, which was funded by the National Science Foundation through the Massachusetts Institute of Technology.

In recognition of his many contributions to the field of geography and especially geographic education, the National Council for Geographic Education conferred on him its Distinguished Service Award in 1973. In 1979 he received an honors award from the Association of American Geographers.

Earl E. Lackey

Earl Emmet Lackey
May 1, 1878 - 1960s?

Earl Lackey was born in a log house near Arlington, Ohio, and moved to Nebraska with his family at the age of twelve. His college education began at Peru State College in Nebraska. He transferred to the University of Nebraska, where he received his A.B. and A.M. degrees. He received his Ph.D. from Columbia University in 1933.

Lackey began his teaching career in a rural school near the village of Western, Nebraska (1897). In 1904 he was appointed Superintendent of Schools in Blue Springs, Nebraska. In 1911 he was appointed head of the Department of Geography and Agriculture at State Teachers College, Wayne, Nebraska. In 1919, following World War I, he taught the American Expeditionary Forces in Beaume, France. In 1924 he moved to the Geography Department, University of Nebraska, Lincoln, where he served until 1949, when he was appointed emeritus professor. During his tenure he served several variable terms as head of the department. For two years (1948-50) he was professor of geography at the University of Tennessee, and was visiting professor in summer schools at the University of Tulsa, Western Reserve University, and the University of Kansas.

Professor Lackey rendered valuable service to the National Council for Geographic Education. He served as secretary for four years (1929-33) and in 1938 served as president. He was associate editor of the *Journal of Geography* from 1929 to 1949. Lackey, along with George J. Miller, designed the first Distinguished Service Award, which was presented to William Morris Davis in 1932. Lackey was also active and influential in the Nebraska Council for Geographic Education while at the University of Nebraska.

In 1950, when he was seventy-two years old, Professor Lackey accepted a position as consultant with the Office of the Quartermaster General; for fifteen years he was employed by the Army on a full-time basis in climatological research. He was recognized for sustained superior service in 1962 and was granted a Special Service Award in 1965 for his contribution to climatology as a member of the Earth Science Division, U.S. Army, Natick Laboratories. In 1967 he retired.

Thus, after half a century of teaching, counseling and writing textbooks, Professor Lackey demonstrated the functional nature of geographical knowledge and techniques, and proved that valuable research can be contributed even after retirement.

Professor Lackey's list of books, articles, papers, and addresses contains approximately 100 titles. The Distinguished Service Award Commit-

Cotton Mather

tee, in a published citation (Cunningham, 1967:522), lists his chief educational publications: *Studies in Principles of Geography, Elementary Geography* (with Richard Dodge), *Advanced Geography* (with Richard Dodge), and *Regions and Nations* (with Esther Anderson). His geographic specialities were weather and climate, geography in education, and the regional geography of continents.

Cotton Mather
January 3, 1918

Cotton Mather was born on a farm in Iowa, in the tiny Quaker hamlet of Springdale, five miles east of Herbert Hoover's hometown. Cotton spent much of the first twenty years of his life in the Midwest and California in circumstances that made a distinct imprint on the person he was to become.

If Cotton Mather had been born in Iowa City to the west, the outcome might have been different. The difference did not lie in the land itself, but in the uniquely homogeneous community and family traditions of education, music, art, and travel in which Cotton Mather grew up. Here were folks with stern morals and a low tolerance for money loosely spent, and conversely, a high regard for honesty and for knowledge put to wise and practical use.

After his early schooling in Iowa and California, Mather entered the University of Iowa in 1935. Later he transferred to the University of Illinois, where he received his A.B. degree in 1940 and his M.S. in 1941.

After receiving his masters degree, Mather served as map editor in the Army Map Service at Washington, D.C. for one year. He later worked as a research analyst in the Office of Strategic Services, Washington, D.C. during the Second World War, interrupting his graduate work at the University of Wisconsin. He received his Ph.D. from the University of Wisconsin in 1950.

Cotton Mather launched his teaching career at the University of Wisconsin in 1946 as a geography instructor, later moving to the University of Georgia in 1947. His scholarly habits and intellectual aspirations served him well during the heady atmosphere of geography in the 1950s. The full panoply of Cotton Mather's wide-ranging achievements, particularly in field seminars, began to unfold. The results of his field studies began to appear in journals such as *Economic Geography* (1950),

Geographical Review (1954), *Land Economics* (1956), *Landscape* (1957), and many others.

In 1952 Cotton Mather, along with Fraser Hart, directed a field trip across the Southeastern United States for delegates attending the International Geographical Congress in Washington, D.C. Few experiences in geography can be as delightful and rewarding as being on a field trip with Cotton Mather.

In 1957 Mather moved to the University of Minnesota as professor of geography. He also served a term as department chair, supervised the graduate work of a score of students, and served as mentor to many others. After twenty-six years at the University of Minnesota, he retired in 1983. During this period he held fourteen visiting professorships at leading American, Canadian, and overseas universities.

Cotton Mather has published widely in leading European and American journals. Many of his writings are required reading in graduate seminar courses throughout the world.

Cotton Mather is currently president of the oldest geographical society in the American Southwest and devotes much of his time to fostering the discovery and dissemination of geographical knowledge through field seminars in the West.

As a geographer, Mather has always quickly responded to places of natural and cultural beauty. In the American West, Latin America, and in India he did not have to pick and choose; breathtaking natural and cultural beauty was everywhere to study and explore. Of particular fascination on Mather's first trip to India was the Vale of Kashmir and the Mount Everest region. In India another kind of beauty caught Cotton Mather's eye — folk art. Truly captivated, he claimed the subject of Indian paintings as his own, learning all he could about them. Later, he coauthored research papers on the geography of folk art.

Cotton Mather has taught a whole generation of geographers about cultural and rural landscapes, stressing the importance of getting out into the field to keenly observe the surroundings and speak with the people who live there. Mather is dedicated to the importance of excellence in teaching and to an extremely flexible graduate program in which students are treated as mature adults. Cotton has found many ways of serving the geographic profession, the community, the state, and the nation.

He is chairman of the Board of the Center for American Places, which fosters publications on landscapes in association with the New Mexico Geographical Society. Along with George Thompson, president of the Center for American Places, Mather is busy planning the publication of books on the North American environment, regions, landscapes and cities.

We have a splendid role model in Cotton Mather, who is an outstanding example of what a great geographer-educator should be.

Alfred H. L. Meyer

Alfred Herman Ludwig Meyer
February 27, 1893 - February 27, 1988

Alfred Meyer was born and raised in Venedy, a village in Washington County in southeastern Illinois, twenty-eight miles southwest of Centralia in an agricultural and bituminous coal mining area. Professor Meyer was a midwesterner by birth and education. For nearly half a century he spent his academic life here. Through careful research, diligent publications, time-consuming leadership and faithful service, he strove to promote better geographic education, thus earning state, national, and international recognition.

Alfred attended the Venedy St. Salvador Lutheran one-room school during the mornings (1900-08) and the Venedy public one-room school during the afternoons (1900-09), graduating from the ninth grade in 1910. He attended secondary schools at Southern Illinois Normal University (now Southern Illinois University at Carbondale) during the summers of 1911 and 1912, Eastern Illinois Normal (now Eastern Illinois University at Charleston) in the summer of 1913, and McKendree College (Academy), Lebanon, Illinois, during 1914-16.

Meyer started as a freshman at McKendree College (1916-17), then transferred to the University of Illinois (1917-18 and 1919-21), where he earned an A.B. in geology in 1921 and an A.M. in geology in 1923. After attending the University of Michigan during the summers of 1928, 1929 and 1930, and spending a year's residency there in 1932-33, he earned his Ph.D. in geography in 1934. He also did graduate work at Northwestern during the summer of 1927 and spent several summers at the University of Chicago studying geology, education, and biology. His doctoral dissertation, "The Kankakee Marsh of Northern Indiana and Illinois," was published in part in the *Proceedings* of the Michigan Academy of Science (Vol. 21, 1935). Meyer maintained a research interest in that area, and perhaps his greatest research contributions are about the Calumet region of Indiana-Illinois, where he took his students on field trips.

Meyer's teaching career started in Linneman — in a one-room rural school six miles south of Venedy. He taught next in Venedy's one-room public school (1911-14). After securing his masters degree, he taught at the secondary level in the Englewood Luther Institute in Chicago during 1925. Professor Meyer's method of alternating teaching with going to college was a common practice in the early half of the twentieth century.

In 1926 Meyer joined the faculty of Valparaiso University as an instructor in geology and zoology. In 1931, as assistant professor and head, he helped establish the Department of Geography and Geology. He was made a professor in 1942 and later became a distinguished service professor. For over thirty years his leadership and influence flowed outward from the department. His undergraduate department educated thousands of geography, physical science, and social science teachers for both elementary and secondary schools. The department also provided majors with such superior subject-matter backgrounds and teaching know-how that this private, church-related college has had more of its graduates take advanced degrees in geography than any other undergraduate college in Indiana (Barton, 1969:560).

In addition to his heavy departmental and college administrative duties, Meyer compiled an impressive publication record based on both library and field research. Between 1932 and 1969 he authored or coauthored over three dozen articles and papers and was coauthor or contributing author of six books and monographs. Eight of these articles were published in the *Journal of Geography*; the remainder in ten other scientific and educational publications. Perhaps his best known book, coauthored by a departmental colleague, is *Geography in World Society — a Conceptual Approach.*

Meyer made major contributions to state and national organizations. He served as a member of the Executive Board and was president of the following organizations: National Council for Geographic Education (1947), City Planning Commission of Valparaiso (1948, 1949, 1950), Indiana Academy of Science (1953), and the Indiana Academy of Social Sciences (1957). He also served as director of the Chicago Regional Planning Association and as a member of the Executive Board of the Association of American Geographers.

Professor Meyer's superior teaching ability, sound research scholarship, and national leadership was recognized throughout the United States. He taught as a visiting professor at eleven universities and with the Valparaiso University overseas program in Reutlingen, Germany, in 1968.

George J. Miller

George J. Miller
February 7, 1880 - July 23, 1973

George J. Miller, founder of the National Council for Geographic Education and editor of the *Journal of Geography* for over thirty years, was born in Sturgis, Michigan. As a youth he attended a public grade school near Constantine, Michigan. After graduating from Constantine High School, George taught in a one-room rural school with grade levels from first to eighth inclusive. He disliked the rote learning and the memorization of encyclopedic facts which he had been taught in school, but he used some of these same methods in his teaching because he knew of no other techniques or methods to use. Thinking there must be a better way to teach, Miller attended Michigan Normal School (now Eastern Michigan University) during 1898-90 and graduated with a two-year diploma (Barton, 1974:53-59).

As did many other young people in his day, Miller was able to earn a living and pay for his university education by alternating teaching with periods of attending college. After earning his two-year diploma, Miller taught in the elementary school at Burr Oak, Michigan (1900-01); was superintendent of the elementary school in Colon, Michigan (1901-03); and was principal of the Woodward School, Kalamazoo (1903-07).

Poor geography teaching influenced Miller to attend the University of Chicago during the summers of 1905 and 1906. After seven years of teaching public school (1900-07), Miller returned to the University of Chicago. He became an instructor in the high school affiliated with the University of Chicago's School of Education in 1908, but continued to enroll in graduate geography classes. He received his M.S. in June 1909, and by the spring of 1913 he had nearly completed the work for his doctorate in geography.

By June 1913 Miller had one manuscript published and five or six accepted for publication. In 1911 he had published the "Establishment of Michigan's Boundaries: A study in Historical Geography" in the *Bulletin of the American Geographical Society*. Two years later the same magazine published "Some Geographic Influences in the Settlement of Michigan and the Distribution of Its Population." Another article, "A High School Course in Field Geography," appeared in *Nature Study Review* in 1913. Early in 1914 the *Journal of Geography* published "Essentials of Modern Geography and Criteria for Their Determination." These titles indicate Miller's interest in both research and geographic education, which continued throughout his life.

Because of his diverse teaching experience, publications and dynamic personality, Miller was invited in 1913 to become head of the Department of Geography at Mankato Teachers College (now Mankato State College) in Minnesota, where he stayed for fifty-five years. Recognizing that geography teachers throughout the U.S. were all facing similar instructional problems, Miller felt that a national organization of geography teachers was needed.

Miller's dream of establishing the National Council of Geography Teachers became a reality in December 1915. With selfless, tireless energy and enthusiasm, he not only brought the Council into existence but implemented, guided, and nourished its growth and accomplishments during its first thirty-five years (1916-50). He served as national secretary from 1916 to 1929; his dynamic energy and wise leadership provided the glue which held the organization together. Two years after relinquishing his secretarial duties, Miller was elected second vice president in 1931, first vice president in 1932, and president in 1933.

In 1920 Miller succeeded Isaiah Bowman as editor of the *Journal of Geography*; he remained editor for thirty years. Miller helped write the NCGT's constitution and, as editor, initiated the idea of the NCGT publishing the Professional Papers Series. Miller was the organization's historian and reference man. In addition to serving on most of the committees, he chaired many of them because he was well informed, contributed innovative ideas, and was willing to give generously of his time and talents. He and Professor Earl E. Lackey drew up the criteria governing the selection of the persons on whom the Distinguished Service Award was to be conferred.

In addition to serving as editor of the *Journal of Geography* in some capacity from 1920 to 1978 (Barton, 1973:54), Miller also contributed to and edited the five volumes of the NCGT's Geographic Education Series. He, along with Zoe A. Thralls and Marguerite Utley, produced Geography Test Form A for Grade 4, published by McKnight and McKnight, Normal, Illionis (1940). Before becoming editor of the *Journal of Geography,* Miller edited *School Progress*. He, along with Almon E. Parkins, Mark Jefferson and three others, edited the American Geographers Series of political wall maps.

Miller's first book, *Geography of Minnesota,* was published in 1923. He coauthored with Almon E. Parkins a college text, *Geography of North America.* After Professor Parkins's death, Miller and Bart Hudgins published the third revision of *Geography of North America.* Miller also contributed chapters to four multi-authored books and published forty-five articles.

In 1938 the National Council of Geography Teachers Distinguished

Service Award was conferred on Professor George J. Miller. In 1975 the Executive Board voted to change the name of the award to the George J. Miller Award for Distinguished Service "to perpetuate the name of George J. Miller, who more than anyone else, was responsible for the NCGE and the *Journal of Geography* (Warman, 1975:508-512).

John W. Morris

John Wesley Morris
November 14, 1907 - September 8, 1982

John Wesley Morris was born in Billings, Oklahoma. After graduating from high school in 1924, he alternated teaching with university studies. He began his college career at the University of Oklahoma, where he received a B.S. in 1930. In 1927-28 he became principal of a small grade school at the age of nineteen; three years later he became principal of a high school. In 1931 he started teaching at Seminole High School and Junior College in Oklahoma, remaining there until 1938, when he went to George Peabody College for Teachers to work toward his doctorate.

Professor Morris earned his M.S. degree at Oklahoma A. & M. College (now Oklahoma State University) in 1934, and received his Ph.D. in 1941. He served five years (1939-42 and 1946-48) at Southeastern State College, Durant, Oklahoma. In 1948 he was invited to join the Department of Geography at the University of Oklahoma, where he was associate professor, professor, then chairman of the department (1963-73). He was also associate director of the Oklahoma Center for Urban and Regional Studies (1953-73).

Professor Morris's public teaching career was interrupted by World War II. His military career actually started while he was in high school, when he joined Battery F of the 189th Field Artillery. In January 1942, while he was teaching a class at the University of Oklahoma, the university president interrupted the class to show him a telegram from the Secretary of the Navy, asking if Morris could be relieved to teach meteorology to prospective student aviators. Morris was first sent to Dallas as a civilian to help establish a Navy weather station; when it was completed he was transferred to Norman, Oklahoma. In 1942 he was commissioned as a lieutenant. After a year he was transferred to North Carolina and from there was sent to a photo-interpretation school in Washington.

Admiral F.C. Sherman visited the school, however, and selected Morris for his staff. Morris spent the rest of the War in the Pacific as a photo-interpreter and geographer. He was involved in the Philippine invasion and in the South China Sea, Iwo Jima, and Okinawa conflicts, while on the staff of Admiral Sherman, Commanding Carrier Division. In 1950-51, when Sherman was Chief of Staff, Morris was called back into service and was sent to Europe. He served as photo-interpreter and geographer on the staff of Admiral Gardner, who was commander of the Mediterrean fleet. Morris eventually became a commander. He retired from active duty in the U.S.N.R. in 1955 but remained a reservist (letter from Morris to Barton, May 5 1982).

In spite of a triple by-pass operation in 1974 and four operations for cancer after 1977, Professor Morris continued to do research. He edited four books for the Oklahoma Historical Society and also edited the *Historical Atlas of Oklahoma,* wrote and published *Ghost Towns of Oklahoma,* and coauthored a book entitled *Oklahoma Homes: Past and Present.* At the time of his death, Morris was working on a manuscript on the history of the Greater Seminole area. The *Historical Atlas of the Indians of the United States,* which he considered his biggest project, was completed before his death.

Interspersed with and following his government service, Morris spent the summers as a visiting professor at the University of Florida (1957 and 1958), Western Washington College (1962), Michigan State University (1969), University of British Columbia (1971), and the University of Wyoming (1973). During the summer of 1956 he served as an intelligence expert in the Department of the Army.

Professor Morris was a long-time member of the National Council for Geographic Education. He served as secretary (1955-58), coordinator of the Oklahoma Chapter, as associate editor of the *Journal of Geography* (1960-66), and as Director of Publications (1955-63). He became first vice-president of the NCGE in 1959 and president in 1960. He and Mary, his wife and librarian, prepared the *Index to the Journal of Geography, 1922-1956.* Morris also served as the National Council for Geographic Education representative to the International Geographical Union from about 1965 to 1973.

Professor Morris received many fellowships, scholarships, and awards. He was granted fellowships at George Peabody College for Teachers during the summers of 1935 and 1936 and for the school year 1938-39. He was awarded a Navy research grant for 1952-54 and a Ford Foundation grant during 1958-59, as well as numerous grants from the University of Oklahoma Faculty Research Fund. In 1965 he received the 75th Anniversary Distinguished Former Student Award from Central State University in Oklahoma. In 1966 he was awarded the National Council for Geographic Education Distinguished Service Award. In 1978 the Oklahoma Academy of Science selected Morris as the Outstanding Scientist of Oklahoma, and in 1982 the Oklahoma Heritage Association gave Morris the Draper Award for service to the state. His book, *Ghost Towns of Oklahoma,* was given an award by the American Association of State and Local History and was made into a mini-series by an Oklahoma City television station.

Professor Morris also served the Association of American Geographers as chairman of the Southwest Division several times. He chaired the geography section of both the Oklahoma Education Association and the Oklahoma Academy of Science. He also joined the Southwestern Social Science Association and served as chairman of the geography section, and as vice-president, first vice-president, and president.

Edith P. Parker

Edith Putnam Parker
June 7, 1886 - October 10, 1961

Edith Parker rendered distinguished service and exerted national and international leadership for nearly half a century in the field of geography in education through her inspired teaching, scholarly publications, devotion to the continued development of in-service teachers, and pioneer research in geographic education (Barton, 1950:49-50).

Edith Parker was born in Rochester, Illinois, the daughter of Homer D. and Margaret Elizabeth Parker. From 1905 to 1911 she alternated college study with teaching. She attended Wellesley College and Teachers College, Columbia University, for a time but returned to her home state and earned a Ph.B. at the University of Chicago in 1913. She started teaching in the laboratory school of the University of Chicago, and from 1924 through 1939 was assistant professor of geography and education. During this time she developed many special education courses, one of which was "The Teaching of Geography in Normal Schools." She also developed "A Tentative Geography Program for Teachers Colleges" (Phillips, 1964:183).

Early in her academic career, Parker seriously considered becoming an English and music teacher. Fortunately for the geography profession, during her early years at the University of Chicago she became associated with two great geographers — Harland Barrows and Rollin D. Salisbury — who influenced her to specialize in geography (Kemter, 1962:222). Although she taught in the laboratory school of the University, Parker continued with her formal education, receiving an M.S. from the University of Chicago in 1922.

Between 1920 and 1927 Parker and Barrows coauthored a series of elementary school geography textbooks, which were innovative in many ways and broke from the tradition of the two-cycle sequence of elementary presentation. The late Margaret Parker, a distinguished geographer at Wellesley College and sister of Edith, lent a helping hand with the third book, *Europe and Asia*, and fourth book, *Southern Lands*, in the series. In 1933 Parker published *Developing the Science of Teaching Geography*. With the help of Clarence Sorenson, the elementary textbooks were revised between 1946 and 1958.

Although Parker was a member of both the Association of American Geographers and the National Council for Geographic Education, she was most active in the latter organization and served as its president in 1934-35. While she was president and for y ars afterwards, she advocated the expansion of the National Council so as to make it a more scientific organization and not just a society (Phillips, 1964:183). Many of the papers she presented at the annual meetings of the National Council were published in the *Journal of Geography*. Her articles also appeared in many other journals published by other professional organizations, such as the National Education Association and the National Council for the Social Studies.

In addition to being a prolific author, Parker was editorial consultant for United Educator. For more than two decades she was editor-designer for the Weber-Costello Company, assisting in the manufacture of reality maps, globes, and life and latitude charts. She was an educational collaborator for Coronet Instruction Films and helped produce a sound motion picture on Western Europe. She also served as associate editor of the *Journal of Geography* for about two decades before resigning in 1954.

Parker exerted leadership through active participation in organizations which strived for better teaching in geography. Many geographers know of her work in the preparation of the *Thirty-Second Yearbook of the National Society for the Study of Education on the Teaching of Geography.*

Her contributions to Part IV of the *Report of the Commission on Social Studies* and to the *Nineteenth Yearbook of the National Council for the Social Studies on Geographic Approaches to Social Education* were milestones in educational geography.

A constant, steady endeavor to advance the frontiers of research in geographic education was one of Parker's lifelong goals. She established an entirely new division of geographic research, namely the geography of education. In her seminars she directed student research at the doctorate level, and three doctoral dissertations were completed under her direction (Phillips, 1964:215).

Mary Viola Phillips, in her dissertation, *Edith Putnam Parker: Her Work and Contributions,* writes:

> The above assortment of material, and the opinion of her contemporaries, provided strong evidence that she was recognized as one of the foremost leaders in geographic education and creative teachers of geography in this country during the first half of the twentieth century. During her teaching career, she carried out investigations and experimentations related to the curriculum in geography and to the preparation of geography teachers. Her pioneering efforts did much to promote the growth of the content, method, and theory of geography (Phillips, 1964:2).

During Parker's lifetime she received numerous awards and honors, including the National Council for Geography Teachers Distinguished Service Award in 1949 and the Chicago Geographic Society's honorary award in 1959 for "contributions to geographic education through the preparation of teaching materials and the training of teachers in geography." During 1962-63, the Chicago-based Society of Women Geographers established the Edith P. Parker and Edna E. Eisen Fellowship in Geography at the University of Chicago. Zoe A. Thralls, former professor and head of the Geography Department at the University of Pittsburgh, dedicated her textbook, *The Teaching of Geography,* to Parker; and Lucia Carolyn Harrison, professor of geography at Western Michigan University, dedicated her book, *Sun, Earth, Time and Man,* to Parker.

Mary Viola Phillips

Mary Viola Phillips
January 12, 1913 - September 3, 1978

Phillips was born in Vandergrift in west central Pennsylvania, twenty-four miles northeast of Pittsburgh. She received her B.S. from the University of Pittsburgh in 1936. Phillips interrupted her college career to teach for awhile in the New Kensington, Pennsylvania, grade school. She soon transferred to the New Kensington High School, where she developed the geography curriculum that became recognized not only in Pennsylvania but nationwide after her seven articles on the content of the course were published in the *Journal of Geography* during 1945-48.

She returned to the University of Pittsburgh and received her M.A. in 1950. She was a visiting professor at the University of Pittsburgh during the spring semester of 1954 and the summers of 1954, 1955 and 1956. She was an instructor of geography during 1957-58 and 1958-59; during the next four years (1959-63), she was assistant professor. In January 1963 Phillips was appointed assistant professor of education, and in 1963-64 she received a joint appointment as assistant professor of geography and education. Professor Phillips earned her doctorate (Ed.D.) from Columbia University in 1964; her dissertation was on the work and contributions of Edith Putnam Parker (Anderzhon, 1979:6).

Phillips served both the National Council for Geographic Education and the Pennsylvania Council for Geographic Education with enthusiasm and distinction. She coordinated her contributions to both with masterful presentations in lectures and publications for more than three decades. Colleagues and students still use her publications on geographic education.

Phillips published fourteen articles in the *Journal of Geography* — six before 1948. These six were printed in *Geography in the High School-1949.* Two of the fourteen articles received national awards from the National Council for Geographic Education for the best article published in the *Journal of Geography* on techniques in teaching geography at the secondary school level (1952 and 1958).

Phillips became the state chairperson for the development of the *Pennsylvania Curriculum Bulletin for Geography*, Department of Public Instruction, Harrisburg, Pennsylvania. From the beginning of her career she was active in the Pennsylvania Council for Geographic Education and was its president in 1954-55 (letter to Barton from Dr. Andrew Petor, husband of Mary Viola Phillips, September 7, 1982). She served on the Curriculum Committee and as the PCGE's state historian. She also served as geographer on the Earth Science Committee, Department of Public Instruction, Harrisburg (1958-59).

Phillips was a life member of the NCGE and served on numerous com-

George T. Renner, Jr.

mittees and on the Executive Board. She served as second vice-president, first vice-president, and president in 1958 (*Journal of Geography*, 1977 [76]:276). She was associate editor of the *Journal of Geography* during 1955-65, and director of publicity in 1954-56.

In addition to articles published in the *Journal of Geography*, she contributed to NCGE publications: "Suggested Sequence for Geographic Learning;" Chapter 3 in *Curriculum Guide for Geographic Education* (1963); and "Europe," Chapter 17 in *Methods of Geographic Education* (1968).

From 1959 to 1963 Professor Phillips represented the NCGE on the Joint Committee on Education of the Association of American Geographers while working on the NCGE High School Geography Project. In 1960 her book, *World Geography for High Schools*, was published by Holt, Rinehart, and Winston. During her extensive travels abroad — lecturing in schools, libraries, museums, government centers and industries — she assembled an outstanding collection of slides for classrooom use.

Professor Phillips received many honors. She was elected to the Society of Women Geographers and was a member of Delta Kappa Gamma. In 1976 she received the PCGE Distinguished Service Award, and in 1977 received the George J. Miller Distinguished Service Award.

Her death left a void in the leadership of geographic education at the University of Pittsburgh and in the National Council for Geographic Education.

George Thomas Renner, Jr.
Early 1900s? - October 14, 1955

George T. Renner was born at the turn of the twentieth century at a time when technology was beginning to reduce the distance between nations and bring them closer together. Renner spent a lifetime analyzing, describing, and popularizing the air age to university students and to the public as he attempted to revitalize the field of geography.

Professor Renner's versatility started during his early youth in Winfield, Kansas. His father, a Methodist minister, encouraged his son to study the classics and read such volumes as Gibbon's *Rise and Fall of the Roman Empire*. In his early training, young Renner read books on history, economics, anthropology, ethnology, geology, and English. A sense of curiosity and scientific interest enriched his background and gave evidence of originality and nonconformity in his thinking. During his teenage years, Renner constructed a wireless telegraphic station and also attempted a flight in a balloon.

Renner received a B.A. degree in 1922. After initial graduate work in geography at the University of Chicago, he transferred to Columbia University, where he majored in the social sciences and received an M.A. in 1924 and a Ph.D. in 1927. In 1936 he joined the faculty of the Teachers College, Columbia University, where he became a full professor in 1939 (Tuthill, 1956:93-95).

Renner was associate editor of the *Journal of Geography,* and for a number of years was geographical editor for a New York publishing house, where he worked to develop symposium-type textbooks. *Global Geography,* published in 1944 by George Renner and Associates, was a pioneer effort among modern geography texts. Renner was convinced that a number of scholars — under the direction of a supervising and contributing editor — could produce a superior textbook (Pearcy, 1958:244-249).

In some respects Professor Renner was ahead of his time; his ideas were advanced beyond the norm of the day. He not only developed new approaches and patterns of ideas, but presented them in a startling fashion that jarred his listeners and readers. A number of his colleagues considered some of his lectures and publications sensational rather than scientific. Regardless, Renner's thought-provoking ideas often stimulated lively discussions, which resulted in new, constructive points of view. Many times his predictions came true. Before World War II he visualized the potentials of aeronautical science and attempted to introduce a polar projection on which to highlight the Great Circle routes across the North Pole. However, it was only in the latter part of World War II, when the airplane came into popular use on long distance flights, that his innovation became generally accepted.

Professor Renner published articles on the economic, political, physical, social, and cartographic phases of geography. A brief analysis of Renner's publications (Pearcy, 1958:246-249) substantiates his wide interests. Many of the topics are regional in nature, including detailed studies on individual countries and regions in Latin America, Africa, Europe, Asia, and many parts of the United States. Many of his articles on geography and education were widely republished throughout the U.S. and Europe. Renner was considered an expert in military geography by the U.S. Government, which he served as a consultant. He also specialized in land planning and various other aspects of conservation.

In contrast with many other geographers who were awarded the Distinguished Service Award, Renner did not combine foreign travel with his wide range of regional studies and writings. He nevertheless maintained a deep appreciation for the major problems of most of the world. He not only traveled in various regions of the U.S., but also served as visiting professor in various states. Before 1936 — when he joined Columbia University permanently as a professor — he had taught at the Columbia School of Business, University of Washington, University of Southern California, Stanford University, University of Minnesota, George Washington University, California State College at Fresno, San Jose (California) State College, Utah State Agricultural College, and at Stetson University in Florida.

As an avid and prolific writer, Renner wrote for both students and the general public. Believing in popularization as a method for stimulating public interest, he became a contributor to several of the more popular magazines. He expended enormous energy to fulfill the need for wider geographic appreciation and understanding in American education. The 1956 National Council of Geography Teachers Distinguished Service Award was conferred upon George Thomas Renner posthumously.

Ina Cullom Robertson

Ina Cullom Robertson
June 21, 1899 - November 11, 1975

Ina Cullom Robertson was born and raised in Frankfort, Illinois, where she graduated from Du Quoin Township High School. She attended De Pauw University, and was awarded a B.S. from the University of Illinois in 1921. She received a scholarship from Clark University to work toward a masters degree, which she received in 1924. She taught in rural elementary schools, high schools and colleges, both in the U. S. and in England. In 1938-39 she was an exchange teacher at the Tadcaster Grammar School, Yorkshire, England.

In 1921 Robertson joined the faculty of the State Teachers College, Valley city, North Dakota, to fill a temporary vacant position. Due to an accident, the "on-leave" professor did not return, and Robertson remained there for thirty-seven years. She dedicated her professional career to the development of the geography department, which increased progressively in strength, efficiency, and in its enrollment under her capable leadership (Miller, 1966:15-23).

In 1925 Robertson began the first student field trips at the college. National attention was focused on the Geography Department when its students, traveling under the supervision of Robertson, "covered all parts of the United States at a time when long-distance excursions were a rarity" (Miller, 1966:19). By 1932 the departmental program developed by Robertson was firmly established and the number of geography majors was growing steadily.

After retiring, Robertson was hired by the Anthony Tour Agency, where she worked from 1958 tc 1961. She remained a life-long student by engaging in personal travel, supervising student field trips to different areas of the United States, teaching overseas, and serving as an official delegate for one of the several geographic organizations to which she belonged. These activities gave her the opportunity to travel to countries in Eurasia, Africa, South America, and North America (Miller, 1963:423).

Robertson lectured at many county and state institutes, both in North Dakota and in the adjacent states. She gave three lectures at the Second Conference on Geography Teaching at Montana State College, Bozeman, in 1955; all three were selected for publication in the *Proceedings* of the meeting.

In spite of her innovativeness in well-planned and supervised student field trips and in covering an extensive public speaking circuit, perhaps Robertson's greatest contribution to geography was her inspirational teaching. She encouraged many of her students to major in geography and to continue their study in graduate school. Some of her students

Neville Scarfe

are now teaching in leading universities.

Robertson was elected as the National Council of Geography Teachers' second vice-president in 1951, secretary in 1952-55, first vice-president in 1956, and president in 1957.

Robertson wrote over a dozen book reviews for the *Journal of Geography* from 1922 to 1956. One of her better known articles, "Ozark Region of Illinois — A Regional Study of Land Utilization," was published in the *Journal of Geography,* 1925 (24):101-111. One of her major studies for the NCGE, "A Survey of the Academic and Professional Preparation of the Critic or Supervisor of Student Teaching in the Field of Geography," was published in the *Journal of Geography;* and was evaluated, selected, and issued as Professional Paper No. 13 in the NCGE's Professional Paper Series. Robertson also coauthored *Geography of North Dakota.*

After retiring from college teaching, Robertson continued to lecture, visit geography departments, and prepare travel booklets as she constantly stressed the value of student travel in geographic education.

In 1963 Ina Robertson's long and selfless service to her profession was recognized when she received the National Council for Geographic Education Distinguished Service Award.

Neville V. Scarfe
March 8, 1908 - early 1980's?

Neville Scarfe, a Canadian, was the first non-U.S. citizen to be awarded the George J. Miller Award for Distinguished Service (1975). Neville was born on a farm in the village of Panfield, near the town of Braintree, in Essex County, England. His early education was obtained in King Edward VI Grammer School, Chelmsford, Essex. He did his undergraduate work at University College, London, where he received an B.A. degree with honors in 1927, a teacher's diploma in 1928, and his M.A. in geography in 1932. Scarfe's masters thesis, "Agricultural Geography of Essex County, England" was completed under the supervision of the world-famous British geographer, Charles B. Fawcett. Scarfe was the top geography student in University College, and received a scholarship to the University of Chicago.

Scarfe's teaching career began with a two-year assignment as Senior

Geography Master at Benrose School, Derby, England. Next he held a four-year position as lecturer in geography at the University of Nottingham, England. For sixteen years (1935-51) he served as head of the Geography Department at the Institute of Education, University of London. During the war years (1939-45), he was also Director of Press Censorship in the Ministry of Information.

Scarfe spent 1948-49 as visiting professor at Syracuse University, New York, where he accepted frequent invitations to lecture at nearby universities. His lectures were well organized, packed with vital innovative content, and enlivened with humor.

Professor Scarfe served as coordinator of the UNESCO International Seminar held at MacDonald College, St. Anne de Bellevue, in July and August of 1950.

During the years 1951 to 1956 he served as professor and dean of education at the University of Manitoba, Winnipeg. In 1956 he was appointed professor and dean of education at the University of British Columbia. During 1952-56 he served as chairman of the International Commission on Teaching Geography.

Scarfe enjoyed summer visiting professorships at McGill University (1951), Clark (1952), Western Washington (1958), and the University of South Pacific (1970 and 1972). In 1968 he was invited by the Vice-Chancellor's Committee of New Zealand to be a distinguished visitor and lecturer at all the universities in New Zealand. During the same year he also visited all of the universities and training colleges in Australia. As a Commonwealth Fellow for the Australian government, he advised on teacher education in each state of Australia. In 1969 he received the Nuffield Traveling Fellowship Award, the British Council Scholar Award, and was invited by the South African Department of Education to advise it on teacher education. In 1970 Scarfe was invited by the Geographic Association of Great Britain to be the keynote speaker at its annual conference held at the London School of Economics.

Scarfe established the Faculty of Education at the University of British Columbia in 1956. For many years it was the largest faculty in Canada and second largest in North America. The enrollment in 1956 was 900 undergraduates, but by 1970 the enrollment had increased to 3,400 undergraduates and 600 graduates. He also designed a new education building which, since 1973, has been called the Neville V. Scarfe Building.

Scarfe was a skillful and prolific writer. In addition to such books as *A New Geography of Canada* and *The Land of Britain*, which were published in 1963, he wrote numerous articles for journals and periodicals in most English-speaking nations. He published his first article in *Scottish Geographical Magazine* in 1935; by 1982 he had published 167 articles, pamphlets and books. Of these, ninety-three articles dealt with educational theory and practice; some scholars might classify them as articles on the philosophy and psychology of teacher training. The strictly geographical books and articles numbered fifty-eight. The remaining sixteen were on various aspects of early childhood education. Scarfe's last article, "Historical Perspective in Geographical Education," was published by the International Geographical Union Commission of Education in London.

Although Scarfe reached retirement age in 1973, the University of British Columbia retained his services as a professor for five more years. During those years he also taught geography method courses for secondary schools, conducted two educational tours to China and one to New Zealand, and worked for Early Childhood Education. He also served for five years as chairman of the National Task Force on Children's Play. In February 1981 the Association for Childhood Education International held a dinner in Scarfe's honor at which tributes from all over the world were read.

Scarfe was a member of the National Council for Geographic Education, served as associate editor of the *Journal of Geography* (1958-65), and was elected second vice-president of the National Council for Geographic Education (1963), first vice-president (1964), and president (1965).

Scarfe earned an international reputation as a geographer, administrator, and as an authority on childhood education. He never lost interest in geography when he left the field of academic geography for geographic education. Except for the World War II years, he taught "Methods of Teaching Geography" every year for thirty years (1935-65). His innovative mind, perpetual energy, droll sense of humor, and hearty laugh encouraged both his students and colleagues.

Earl B. Shaw

Earl Bennett Shaw
March 18, 1889 - November 3, 1979

Earl Shaw was born in Monroe, Iowa. Following a tour of duty in the Army during World War I, he resumed his education. He graduated from Washington University, St. Louis, with an A.B. in 1927 and an M.S. in 1929, and was awarded a Ph.D. from Clark University in 1933. Shaw was a visiting professor of geography at the University of Puerto Rico in 1932-33 and professor of geography and head of the department at Massachusetts State College, Worcester, in 1933-59.

During his teaching career and after retiring from Massachusetts State College, Professor Shaw was research associate in geography at Smith College in 1932-33, lecturer in the summer session of 1935 at Clark University, and taught summer school at State College, Buffalo, in 1938 and again in 1942. He was lecturer during a two-month summer session for the World's Federation of Educational Associations on board the ocean liner, Rotterdam, in 1939. He was also invited back to lecture at the University of Puerto Rico during the 1940 summer session.

Shaw also served his country during peacetime. He was head of the geography section, Shrivenham American University, Shrivenham, England, for the U. S. War Department in 1945; was a research member of the Board of Geographic Names; and served in the Office of Strategic Services (Svec, 1960:434).

Once World War II ended, Professor Shaw was again invited to be a visiting lecturer at other universities. He was a lecturer during the summer session at the University of Pittsburgh in 1947; University of British Columbia, 1950; Clark University, 1951; and Columbia University, 1951. He became a Fulbright lecturer in Human Geography, Ibrahim University, Cairo, Egypt, 1952-53. The first summer after he retired from Massachusetts State College, Shaw was a lecturer at Columbia University (1959) and at the University of Delaware during the fall semester. He became professor of geography at Assumption College, Worcester, Massachusetts, in 1960 and taught there until 1972.

During fifty-one years of university teaching, Shaw taught thousands of students. Although he loved to teach and students flocked to his classses, he budgeted his time and energies so he had time for research, publishing, and editing. He was an associate editor of the *Journal of Geography* and *Economic Geography*. He served as vice president of the National Council for Geographic Education, and in 1949 was president. As a member of the Executive Board, he was always dependable. He was chairman of the committee on the up-dating of the National Council as a working organization; this committee's report appeared in the May,

Cora P. Sletten

1952 issue of the *Journal of Geography*.

Shaw published over forty-eight articles in such professional magazines as *Economic Geography, Geographical Review*, and *Science Monthly*. One of his three books, *World Economic Geography*, was published in a paperback edition in Chinese. Shaw was also a collaborator on six additional books, and edited nine educational films.

Professor Shaw traveled and did research in every continent except Antarctica. He was awarded an honorary Doctor of Science degree by Assumption College in 1970 and is listed in *Who's Who in American Education, Who's Who in the East*, and *Who Knows What*. In addition to receiving the National Council for Geographic Education Distinguished Service Award in 1960, Shaw was a recipient of the National Geographic Society's Smith Award.

Cora Pauline Sletten
January 17, 1885 - August 8, 1974

Cora P. Stetten was born in Crookston, Minnesota. She attended the Detroit Lakes elementary school and high school, completing the latter in three years. She received a B.A. from Carleton College, Northfield, Minnesota, in 1906, with a major in chemistry and minors in mathematics and history. After receiving her B.A., Sletten taught natural sciences for sixteen years in the St. Charles and Le Sueur, Minnesota, high schools and in Mitchell High School in South Dakota. She then enrolled in the University of Chicago for graduate work in the geography department. She received her M.S. in 1923, then joined the Department of Geography in Mankota State Teachers College, where she worked with George J. Miller for thirty years, teaching and building the department into one of the best in the state. She taught most of the summers at Mankota State, but was a visiting professor at Iowa State Teachers College, Cedar Falls, the summer of 1937 and at Western Reserve University, Cleveland, Ohio, in the summer of 1938.

Sletten was devoted to the National Council of Geography Teachers and the *Journal of Geography*. She served twenty-three years (1925-48) as an assistant editor of the *Journal of Geography*, contributing two columns — editorial notes and news, and geographical publications — each month to the nine yearly issues.

Sletten served on numerous NCGT committees, but the longest term

Marie Melvina Svec

was spent on the Publication Committee (1938-45). She was president of the NCGT in 1940 and received its Distinguished Service Award in 1953. As a delegate, she represented the NCGT at the International Geographical Congress, Lisbon, Portugal, in the spring of 1949 and at a geographic conference in Switzerland in July 1949.

Sletten became chairman of the Geography and Geology Department at Mankato State College in 1948, and held that position until she retired in 1953. As early as the mid 1930s, she gave a series of talks on geography, which were nationally broadcast on CBS Radio.

M(arie) Melvina Svec
September 15, 1901 - deceased?

M. Melvina Svec was born in Cedar Rapids, Iowa. She earned a B.A. in 1925 and an M.A. in 1930, both at the University of Wisconsin, and took additional work at the University of Chicago in the fall of 1937 and at the London School of Economics and Political Science during 1938.

Svec alternated teaching with studying. She taught in a South Dakota rural school in 1919-20, returning to Cedar Rapids the next year to teach in the Johnson Elementary School. From 1925 to 1929 she taught geography in the Wilson Junior High School. After receiving her M.A. in 1930 she taught during the summer in the Geography Department of the College of Education, Marquette, Michigan; during 1930-47 she taught at State Teachers College, Buffalo; and in the summer of 1943 she taught at the College of Commerce, University of Iowa.

Svec left Buffalo State to go to the State University of New York, Oswego (1947-63). She taught several summer sessions in the Geography Department of George Peabody College between 1949 and 1969, and completed her teaching career (1968-69) at the University of Northern Iowa.

Svec traveled extensively in all of the continents but Antarctica; she especially liked to use river boats and trains so she could have a closer look at the landscape. During a six-month sabbatical from Buffalo Teachers College in 1937-38, she traveled in Europe and the USSR; the sites she visited became an educational experience to share with her students (Phelps, 1962:418).

Zoe Agnes Thralls

Professor Svec also wrote more than thirty articles on both the content and educational phases of geography. Most of these were published in the *Journal of Geography*, but some appeared in other professional journals, such as *Social Studies* and *New York Social Studies*. In addition, she prepared books and workbooks, and as a junior author, worked with Professor Katheryne Thomas Whittemore. After retirement in 1963, Professor Svec wrote articles for the Czech Heritage Foundation, Inc.

The focus of Svec's professional life was to promote geography in schools and colleges. To accomplish this, she gave unstinting support to the Association of American Geographers, the American Geographical Society, the National Council for the Social Studies, and the Society for Women Geographers.

Although she made major contributions and set high standards in a multifacited career, Svec also excelled as a classroom teacher. Her enthusiasm, articulateness, skill in the use of instructional materials, and geographic competence left students vitally interested in the earth and keenly aware of the usefulness and importance of geographic study. The National Council for Geographic Education Distinguished Service Award was conferred upon her in 1962.

Zoe Agnes Thralls
December 6, 1888 - May 23, 1965

Zoe Thralls was born in Butler, Missouri, and attended Springfield Teachers College. She received a masters degree from Columbia University and a Ph.B. from the University of Chicago. She began her teaching career at Indiana State College, Indiana, Pennsylvania, where she was head of the Department of Geography.

In 1927 she went to the University of Pittsburgh to teach education. In 1943 she established a geography department at Pittsburgh, and remained there until her retirement in 1958.

Thralls made noteworthy contributions to the National Council for Geographic Education, serving as a member of the Executive Committee, as first vice president and president. She served as editor of the *Bulletin of the Geography Club of Western Pennsylvania* and as an associate editor of the *Journal of Geography*. In 1951 she received the National Council for Geography Teachers Distinguished Service Award.

Henry J. Warman

In 1950 Thralls was appointed as one of four delegates to attend a six-week UNESCO seminar on the teaching of geography at McDonald College, St. Anne de Bellevue, Quebec, Canada. She was chairperson of the U.S. delegation to the Third General Assembly of the Pan American Institute of Geography and History, and was in great demand as a consultant and panelist at numerous educational conferences.

Thralls was a prolific writer. Most of her publications are listed in *Publications of Leaders in Geographic Education Report I, Journal of Geography,* 1965 (64):169-171. Her textbooks were widely used in the public schools.

Henry John Warman
January 27, 1907 - February 9, 1982

Henry Warman was born in Scranton, Pennsylvania. He earned his B.S. at Pennsylvania State College, Bloomsburg (now Bloomsburg University of Pennsylvania) in 1932 and his M.S. at Temple University in 1938. He held a fellowship at Clark University in 1942-43, and was awarded the Ph.D. in geography at Clark in 1945. Warman became a member of the geography faculty at Clark in 1943; in 1958 he was promoted to the rank of professor.

Professor Warman earned a national reputation as a consultant and authority in geographic education. His work with graduate students and teachers-in-service is well known not only throughout New England but in many sections of the U S. In the Worcester, Massachusetts, schools and in many New England school systems, he was a valued consultant in geographic education and curriculum planning.

Warman published more than thirty articles and papers. About half of them were published in the *Journal of Geography;* the rest were published in other professional journals and magazines. He was associate editor of the *Journal of Geography* from 1953 to 1965. He also authored *Geography — Backgrounds, Techniques, and Prospects; Geography in Human Affairs;* and *Major Concepts in Geographic Education.*

Professor Warman always took an active part in the affairs and problems of the NCGE. In 1954 he served as president; he also served on the Executive Board on important committees both prior to and after serving as president. Quiet in nature, Warman made sure that assigned tasks were completed (Svec, 1968:504). At numerous annual meetings of the NCGE he presented provocative papers and reports.

Professor Warman's lifelong goal was to help students and the public become aware of the role geography played in their daily lives. His

Joe Russell Whitaker

interest in weather and climate led him to write a series of fourteen articles for the Worcester *Telegram*. In another series, Warman reported on studies in Latin America that were funded by Atwood research grants. His work with the Armed Forces dealt with map interpretation and climate.

Professor Warman also contributed to and participated in international meetings. He was the U.S. representative at the Pan American Conference on Geography and History held in Mexico City in 1955, and at the meeting for the International Geographic Union's Commission on the Teaching of Geography in 1966.

Warman was president of Gamma Theta Upsilon, the professional geographic fraternity, and president of the New England and St. Lawrence Geological Society. He taught during the summer sessions at the University of British Columbia, University of Southern California, and at the University of Colorado. He served eleven years in the Pennsylvania school systems as teacher, coach and athletic director, and was the field coordinator for the High School Geography Project (1962-63).

In 1968 Professor Warman was awarded the National Council for Geographic Education Distinguished Service Award.

Joe Russell Whitaker
January 30, 1900 -

Russell Whitaker was born in Cynthiana, Kentucky, and spent his youth in the Bluegrass, where he had ample opportunity to observe the changing landscape as developed by the westward pioneers.

In 1918 he entered Berea College, where he studied for two years before enrolling at the University of Chicago. Soon after receiving a B.S. degree in 1922, he accepted a graduate teaching assistantship at the University of Wisconsin.

By the time he received his A.M degree, Whitaker had taught four years. He had taught geography, biology, and physics in the Academy (high school) at Berea College from 1917 to 1919 and was a teaching assistant at the University of Wisconsin from 1922-23. With this experience, he was hired to teach in the Department of Geography at George Peabody College for Teachers during the summer of 1923.

Professor Whitaker's career as a full-time college teacher began in 1924 when he became head of the Department of Geography at State Teachers College, Marquette, Michigan (Morris, 1968: 189-191). During the next six years he found time to complete his graduate studies at the University of Chicago, and complete the field work for his doctoral dissertation. He received his Ph.D. in 1930. He was then hired by the University of Wisconsin as an assistant professor and later advanced to associate professor.

Whitaker carried on post-doctoral research in Ontario in the summer of 1935 with a grant-in-aid from the National Research Council. In the summer of 1939 he was visiting professor at the University of Minnesota. The George Peabody College for Teachers invited Whitaker to join its faculty as professor and chairman of the Department of Geography in 1940, where he remained until he retired in June 1968.

Whitaker was a member of the National Research Council for fifteen years (1939-53). During World War II he served as a consultant to the Department of State and the Department of War. In 1947 he was advisor to the U.S. Occupation Forces in Germany. During the second semester of the 1965-66 academic year he was visiting professor at Vanderbilt University.

After retiring in June 1968 after twenty-eight years on the Peabody College faculty, Whitaker and his wife went to Greece, where he taught in the Peabody-in-Athen's program during the summer. He taught the following eleven years at the University of Tennessee, and in the fall of 1979 at Tennessee State University, with which the University of Tennessee had just merged. He retired from teaching when he was one month less than eighty years old.

Although much of his time and energy was devoted to teaching, Professor Whitaker found time for research, writing and editing. He authored sixty articles, which appeared in professional journals and magazines. He was the author of seven books, contributed chapters to eighteen additional books, edited eight other books, and served as editorial consultant on still more books. He also helped design and edit *The World Book Atlas*.

Professor Whitaker has been honored in many ways for his outstanding accomplishments and contributions to the geography profession and to his country. He was elected and served as president of the National Council of Geography Teachers in 1938 and of the Association of American Geographers in 1953. He is one of the select few to hold the presidency in both organizations. In 1955 Professor Whitaker was presented the Distinguished Service Award by the National Council of Geography Teachers. In 1968 he was elected to honorary fellowship in the American Geographical Society in recognition of his "distinguished career and contributions to geographical knowledge." In 1979, at a meeting of the Southeastern Division of the Association of American Geographers, the Division awarded him a citation "in recognition and appreciation of his many contributions to the development and progress of professional geography."

Most of Professor Whitaker's students, who are teaching at all levels from first grade to graduate schools throughout the U.S., agree that Whitaker was at his best in the classroom and as a counselor. After observing his teaching skills, experienced teachers returned to their classrooms with valuable and more skillful teaching techniques.

Gilbert White

Gilbert F. White
November 26, 1911

Gilbert White received his S.B., S.M., and Ph.D. degrees from the University of Chicago. His initial professional employment was as a geographer under Harlan H. Barrows with the Mississippi Valley Committee, National Resources Committee, and the National Resources Planning Board during the 1930s. This was followed by three years at the U.S. Bureau of the Budget in the Executive Office of the President.

During the Second World War, White was relief administrator for the American Friends Service Committee in France and later was secretary of American Relief for India. This was followed by ten years as president of Haverford College. In 1956 he returned to the University of Chicago as professor of geography, and in 1970 became director of the Institute of Behavioral Science and professor of geography at the University of Colorado.

White, who is one of our most admired and respected geographers, has strived consciously to promote geographic education and research. His great and enduring contribution to geography is "Natural Hazards research," of which he is a pioneer. He has served on many scientific missions, in the U.S. and around the world, with a devotion to truthfulness and concern for the interests of ordinary people of the area under review. During 1976-82 White was president of the International Council of Scientific Union's Committee on Problems of the Environment.

Foremost among Gilbert White's many contributions to geographic education is his role as chairman of the Steering Committee for the High School Geography Project (1964-70) and his participation in the discussions preceeding it. White helped marshal the inspiration and enthusiasm of many geographers, thus assuring that the High School Geography Project would flourish.

White served a distinguished term as president of the Association of American Geographers (1961-62), and was chairman of Resources For the Future, Inc. (1974-79).

Apart from his scholarly accomplishments, White's influence has been felt through serving on numerous national and international committees that set policies and trends in areas such as water resources, arid zone research, integrated river development, weather modification, federal flood policy, energy policy, and technology assessment. His numerous books and scientific papers have made a substantial contribution to many aspects of geography. Gilbert White is an exemplary proponent of geography in his contacts with other scientists, planners,

Katheryne T. Whittemore

and civic functionaries.

White's contribution to geographic education and research has been recognized on several occasions. He received the Quantrell Award for Excellence in Undergraduate Teaching in 1967; the Distinguished Service Award of the Association of American Geographers in 1955 and 1974; the Daly Medal of the American Geographical Society, 1971; the Thomas Jefferson Award, University of Colorado, 1973; and the Environmental Award, National Academy of Sciences, 1980. In 1981 the National Council for Geographic Education conferred the George J. Miller Distinguished Service Award on Gilbert White.

Katheryne T. Whittemore
November 25, 1899 - June 23, 1981

Katheryne Thomas Whittemore was born in Minden, Nebraska. Her parents were both educators. After graduating from Lincoln High School in Lincoln, Nebraska, Whittemore attended Cotner College, Bethany, Nebraska, 1916-17; Randolph-Macon Women's College, Lynchburg, Virginia, 1917-18; and received an A.B. degree from Vassar College in 1921, where English was her major and economics and history were minors.

In 1921 Whittemore began her career by teaching English in a Fairbanks, Alaska, high school. She later taught in Waseda, Kotogakuin, and other schools in Tokyo, Japan, and in the YMCA High School, Manila, Philippine Islands. After traveling around the world teaching for two years, she returned to the U.S. and started graduate work at Clark University, where she received an M.A. in geography in 1925.

From 1925 to 1962 Whittemore was on the faculty of State University College, Buffalo, New York, as an instructor of geography. She served as chairman of the Geography Department, and in 1957 was appointed director of the Division of Arts and Sciences, which included all academic departments of the college. She received a Ph.D. in geography from Clark University in 1936. For many years she taught a geography class to elementary level students in addition to college teaching, a policy which gave her an exceptional perspective and understanding in preparing teachers for more effective teaching (Meyer, 1959:458-459). Through this practice, both colleagues and students shared the benefits of Whittemore's resourcefulness in the organization of materials, methods of presentation, and in the inspiration of her teaching.

Although she spent a large share of her time, energy, and talents in teaching and consulting with students and faculty, Whittemore's personal and professional influence was not limited to the classroom. She was curriculum consultant to the New York State Education Department when it was formulating achievement examinations in elementary social studies in 1960, and was a frequent consultant at teachers' workshops in New York State. She also served as editorial consultant for Morran Junior Books.

Professor Whittemore not only lectured to education groups on travel, textbooks, and the teaching of geography but was also a popular lecturer for social groups, such as the Rotary Clubs and Museums of Science. With two colleagues, she wrote and presented a thirteen-week television series, "Learn and Live," for WBEN-TV.

A bibliography of Whittemore's publications which appeared prior to 1965 was published in the *Journal of Geography*, seventeen years before she retired (Barton, 1965:211-213). She continued to do research and publish articles and books after her retirement.

Whittemore was an inspirational teacher because of her vivid descriptions of places she had seen in her extensive travels and her splendid use of visual aids, especially kodachromes. Her travels were well-planned to give a complete geographic picture of a region, industry or country. She used non-touristic methods of travel — horseback, riverboat, schooner, truck, train and car — to get to relatively isolated places. She traveled to all areas of New York, collecting information and taking a large number of kodachromes to make the geography of New York State more exciting in the classroom. She also traveled extensively throughout the United States, Africa, and Europe.

Whittemore's professional and societal affiliations and activities were diverse. Because of her talents and dynamic personality, she was seldom just a member but generally an active, working member of organizations she belonged to.

In addition to receiving the National Council for Geographic Education Distinguished Service Award in 1959, Whittemore received the Susan B. Anthony Award from the Interclub Council of Western New York in 1968, and the President's Distinguished Service Award from the State University of New York College, Buffalo, in 1977. One of her books, *The United States and Canada* (a fifth grade textbook), was selected for exhibit at the Brussels World's Fair. In 1962 she was invited to deliver the commencement address to the State University College graduating class, the first faculty member so honored.

Samuel Van Valkenburg

Samuel Van Valkenburg
September 14, 1891 - April 18, 1976

Samuel Van Valkenburg was born in Leeuwarden, the Netherlands. He studied geography at the University of Utrecht in 1913, at the University of Berlin in 1915-16, and at the University of Zurich in 1913-14 and again from 1916 to 1918. He did post-doctoral work at the universities of Neuchatel and Lausanne, and received an honorary Doctor of Law degree from Clark University in 1971.

Van Valkenburg was a leading student of political geography and a recognized authority on the geography of Europe and on climatology. His research ranged from the geomorphology of the Alps to estimates of the future of the European Common Market.

Van Valkenburg served as a geographer in the Netherland East Indies Government for five years (1921-26) before he became a visiting professor during the second semester of 1926-27 at Clark University and an assistant professor of geography in 1927.

In 1928 he served as associate professor of geography in the College of the City of Detroit (now Wayne State University). In 1932 Van Valkenburg returned to the Graduate School of Geography at Clark University as associate professor of climatology and regional geography and served in this capacity until February 1937, when he was made a professor of geography. In 1946 Van Valkenburg was named director of the Graduate School of Geography, where he served until 1962, when he became professor emeritus.

Van Valkenburg's contributions to geography include several college texts and numerous articles. He was coauthor of *Europe, Elements of Political Geography.* He served as managing editor of *Economic Geography* and participated in the annual meetings of the NCGE and other geographic organizations. He served as vice-president of the Association of American Geographers. He was chairman of the International Geographical Union's Land Use Commission and proposed a World Land Use Survey to the Union at its meeting in Lisbon, Portugal, in 1949. He suggested that the Union undertake a World Land Use Survey — based on work done in Britain — to provide a base for increased food production (Valkenburg, 1968:524-525). He was so devoted to this idea that he made a survey of land use for Western Europe.

During World War II Professor Van Valkenburg was a consultant to the Quartermaster General and a member of his Advisory Board. He was also a consultant in the Office of Strategic Services, and to the European Common Market. In 1961 Professor Van Valkenburg was awarded the National Council for Geographic Education Distinguished Service Award.

Stephen S. Visher

Stephen Sargent Visher
December 15, 1887 - October 25, 1967

Stephen Visher was born in Chicago. His father, the Rev. John Visher, who was working with "underprivileged" people in Chicago in 1887 when Stephen was born, had previously served as a Unitarian missionary with the Indians in Minnesota. Although Visher's earliest memories were of urban scenes, he seemed more typically a vigorous, independent son of the "Middle Border" of the Northern Plains. The family (four sisters and two brothers) moved in 1898 to a farm in Sanborn County, South Dakota, where a diverse physical landscape included lakes, prairies, and scattered groves along the James River. This outdoor laboratory became Visher's classroom when his school and farm work schedule permitted. Here was the opportunity for a wide-ranging observation and collection of native plant and animal life, especially birds. In later life Visher expressed a deep gratitude for his early study of South Dakota birds, a major influence in his life.

A study of the youthful Stephen Visher reveals his early promise, his almost limitless curiosity, his devotion to the discovery of new knowledge, his sharing of new information with others, and his persistence in activities undertaken.

> His limited participation in formal education during his formative childhood years and the absence of extensive and congenial peer-group interaction and associated socialization may account for some of his characteristic personality traits revealed in actions which, though logical, struck some of his colleagues at times as eccentric, in the strict sense of the word as "departing from customary behavior" (Harris, 1968:378-379).

During his early childhood, Visher's attendance at public schools was slight since he was tutored at home by his mother. A keen observer of nature (as a fourteen-year-old youth), he saw a rare rosebreasted grosbeak near his home. Delighted, he searched for someone with whom to share his new knowledge. Excitedly, he spoke to two men he found fishing nearby. One of the men happened to be the editor of the county newspapaper. Intrigued by the boy's knowledge and enthusiasm, the man asked the lad to write a newspaper article on the birds of Sanborn county. A copy of the article was sent to Stephen's Aunt Ellen in Chicago, who was so impressed that she invited Stephen to come live with her and attend high school there.

Visher applied for admission to the high school at the old Lewis Institute (now part of the Illinois Institute of Technology), but was unable to prove that he had graduated from the eighth grade. For a month he pleaded for an opportunity to demonstrate his ability to perform high school work. He brilliantly justified his admission by completing the four-year course in three years and graduating as valedictorian.

After teaching at a rural school in his home county for a year, Visher entered the University of Chicago on a special scholarship, completing the undergraduate course in two and a half years. He was elected to Phi Beta Kappa, Sigma Xi, and to the Student Council.

In the summer of 1907 Visher served as assistant to Professor Henry C. Cowle, who directed a botanical field expedition to southern Alaska. In 1909 Visher received a B.S. degree with a major in botany from the University of Chicago. He spent eight months in 1909 at the Carnegie Institute's Desert Laboratory, Tucson, Arizona. In 1910 he received his M.S. from the University of Chicago, after which he was appointed Assistant State Geologist of South Dakota and an instructor at the University of South Dakota, where he served between 1910 and 1913, and also earned an M.A. degree in zoology. Visher authored major publications on his home territory of South Dakota (see publication list in the *Annals,* 1971:394-406).

In 1913 Visher returned to the University of Chicago on a fellowship to finish his doctorate in geography. His dissertation, *The Geography of South Dakota,* was published by the state. He was awarded a Ph.D. with honors in 1914. The same year he married and did field work in the Caribbean. He was professor of geography at State Normal College, Morehead, Minnesota, for three years. During 1918 he classified public lands in Wyoming and Utah for the U.S. Geological Survey and for a short time did field work in South Carolina for the Bureau of Soils.

With this broad background of training and experience, Visher was appointed assistant professor in the Department of Geology at Indiana University in early 1919. According to one survey, Professor Visher wrote three books on Indiana, at least thirty-three articles of a technical nature, more than fifty articles in semi-technical journals, and forty-five major newspaper articles.

Although he concentrated his research on Indiana, Visher's interest in climate was not pushed aside but was expanded and somewhat redirected. As a result he was awarded a post-doctoral fellowship from Yale University from 1920 to 1922. During the first year he traveled in Europe and worked with Ellsworth Huntington on a major book manuscript dealing with climatic changes; the second year he studied tropical cyclones in the far reaches of the Pacific and did field work in Japan, China, the Philippines, Fiji, Hawaii, and Australia. His report for the Australian Commonwealth Bureau of Meteorology, *Tropical Cyclones of the Pacific,* was published by the Bishop Museum, Honolulu, and later translated into Japanese.

In 1924 *Climatic Laws* appeared, followed by numerous articles on various phases of climate in American and foreign journals. He examined the climate of the state for the Geological Survey of Kentucky. This major interest in climate was captured in two books: *Climate of Indiana* (1944) and *Climatic Atlas of the United States* (1953, revised 1966).

In 1931-32 his work as Acting Geographer of the U.S. Department of State resulted in a publication on foreign place names and political boundaries. The book, *Economic Geography,* was coauthored by W.O. Blanchard.

In the field of conservation, Visher worked at times with the U.S. Soil Conservation Service and the Indiana Geological Survey, and wrote professional articles on the subject. At one time or another he was consultant to the Navy, to the Department of Defense, and to Governor Paul V. McNutt of Indiana.

Visher served as a visiting professor during various summer sessions at Cornell, the universities of British Columbia, Colorado, Pennsylvania, Minnesota, and Montana State.

The National Council for Geographic Education, in presenting its Distinguished Service Award to Professor Visher in 1948, noted that he had contributed more articles to the *Journal of Geography* in the last thirty-five years and had written more reviews than any other person (Read, 1949:41). The citation stressed his willingness to serve on committees, appear on annual programs, and attend most of the annual meetings of the National Council.

In 1943 Visher was awarded a Distinguished Citation by the University of Chicago. In 1945 he received an honorary membership in the National Geographic Society, and in 1949 he received the Outstanding Achievement Award of the Association of American Geographers.

Part 3 The Professionals

Mamie L. Anderzhon

Mamie Louise Anderzhon
July 28, 1906 - June 17, 1988

Mamie Anderzhon was born and raised on a farm between Shenandoah and Farragut, Iowa. Her formal education extended from the completion of eight grades in five and a half years in a rural school to many summers of undergraduate work at the University of Chicago and one summer each at McGill, Harvard, and Northwestern universities. She was awarded a bachelors degree in 1938 at the University of Chicago, Division of Social Science, with majors in history and geography; and an M.S. in the Division of Physical Science in 1948, with a major in geography. Edith P. Parker was advisor for Anderzhon's master's thesis, "Geographic Field Work in Community Study."

After graduating from Shenandoah High School in 1923, Anderzhon enrolled in summer school at Nebraska Teachers College in Peru, and received a teachers certificate. Only seventeen, she could not teach in Iowa so she taught in a one-room rural school in Nemaha County, Nebraska, where there were fifty-one students enrolled in grades one through eight. Following a year of teaching in an Iowa rural school, she taught four years in Consolidated Grade School, Farragut, Iowa.

Anderzhon enrolled at Iowa Teachers College, Cedar Falls, in the fall of 1929, where Alison Aitchison encouraged her to teach geography. In August 1930 she received a diploma which qualified her for a position in the Oak Park, Illinois, public schools. She taught geography, history, and civics in Oak Park for three decades except for two interruptions. During 1943-45 she worked for the Washington, D.C. Army Map Service, Cartographic Research Section, Japanese Unit (Barton, 1971:562-563). In 1956-57 she received a Fulbright teaching grant and worked with the All India Council for Secondary Education, assisting classroom teachers in workshops and seminars. The same year she received the 1956 Delta Kappa Gamma fellowship. In 1958 she gave over twenty illustrated lectures on India to students and other educational and civic groups. During the 1950s she directed three workshops at Iowa State Teachers College (now the University of Northern Iowa).

In 1961 Anderzhon was invited to join Indiana University of Pennsylvania's Department of Geography, where she used discovery techniques of teaching at the college level. In 1965 and 1966 she was an evaluator of educational projects for the U.S. Office of Health, Education, and Welfare. In addition to teaching in NDEA Geography Institutes at the University of Hawaii (1965 and 1966) and at Indiana University of Pennsylvania (1966 and 1967), she taught two terms in 1968 at Adelaide Teachers College and Western Teachers College, Adelaide,

Australia. As one of many consultants for the High School Geography Project, she shared with students and teachers in classrooms and workshops the content of the High School Project volumes, which used inquiry and discovery techniques for teaching.

Anderzhon presented papers on the teaching of geography at the International Geographical Congresses in Stockholm (1960), London (1964), and New Delhi (1968). Professor Anderzhon published a number of articles, the first of which appeared in the *Journal of Geography* in 1942. Other articles appeared in such periodicals as the *Journal of Geography, N.E.A. Journal, Grade Teacher, Instructor,* the 1948, 1953 and 1959 Yearbooks of Social Studies, and the National Council for the Geographic Education Curriculum Guide. In 1949 Rand McNally Company published her first edition of *Steps in Map Reading.* By 1970 she had rewritten this first edition, resulting in three volumes which were based on a lifetime of research and experience while working with students at all levels.

In additon to publishing articles and books, "Mamie" (as her friends called her) promoted geography by sharing her color slides with others. She annotated forty of her Australian slides in 1974, with a new printing in 1981, and in 1984 annotated her Republic of Turkey color slides for the National Council for Geographic Education/Great Plains Nebraska (NCGE/GPN) Instructional Slide Library.

Professor Anderzhon served the National Council for Geographic Education and its state chapters well. She served as chairperson of several committees and held elected offices. She served on the Executive Board of the NCGE continuously from 1949 to 1956 and from 1961 to 1965. From 1953 to 1964 she was one of the associate editors of the *Journal of Geography*. She served three years as second vice-president (1952-54), first vice-president (1961), and as president (1962). In 1958 she was asked to chair the National Planning Committee for the second time. She repeatedly appeared in their annual programs. She was a member of the NCGE Curriculum Committee from 1959 until the completion of the Curriculum Guide in 1962-63. Anderzhon was active in the Illinois Geographical Society, Pennsylvania Council for Geography Council (now Pennsylvania Geographical Society), and held continuous memberships in the NCGT/NCGE from 1939 and in the Association of American Geographers from 1947. She was elected to the Society of Woman Geographers in 1953.

Anderzhon attended the International Geographical Congress every four years from 1952 (except 1956), including the 1976 meetings in the USSR. In 1976 she spent two months in the Soviet Union — visiting a number of places by train, bus, plane, and boat. She participated in the IGU "Man and Environment" ten-day symposium, which was held on a cruise ship sailing up the Volga River from Rostov to Kazan.

In 1972 Anderzhon was awarded the honorary rank of emeritus professor at Indiana University of Pennsylvania. Other awards include an Andrew Carnegie Grant for Arctic Study at Magill University Summer Camp, Stanstead, Canada, in 1951; a grant from the National Science Foundation Summer Institute, Northwestern University, in 1962; the Farragut, Iowa, Centennial Special Award for contributions to education in 1971; and the Outstanding Educators of America Award for contributions to higher education and service to the community in 1972. She was made an honorary member of Rand McNally's 25-Year Club for her twenty-five years as continuous author in 1974. She also received a lifetime membership in the Iowa Council for Geographic Education in 1976.

In awarding the National Council for Geographic Education Distinguished Service Award to Mamie, Barton noted that she was:

> "A selfless teacher *par excellence*; an efficient, enthusiastic missionary of geography; a life-long searcher after geographic content and techniques in communicating this knowledge; a lovable world citizen, Mamie Louise Anderzhon" (Barton, 1971:563).

Barry Bishop

Barry Chapman Bishop
January 13, 1932 -

Barry Bishop is one of the strongest advocates of geographic education in America. An outstanding scholar as well as explorer, Bishop was born in Cincinnati, Ohio. He received a B.S. degree in geology from the University of Cincinnati in 1954.

After finishing his undergraduate studies at Cincinnati, Bishop enrolled in the Department of Geography at Northwestern University, where he received an M.A. degree in geography in 1957. As a graduate student at Northwestern, Bishop conducted a geomorphological research project on the Greenland Ice Cap under a contract with the Snow, Ice and Permafrost Research Establishment, U.S. Corps of Engineers. He studied marginal moraine features including glacier-movement, coring, mapping, and aerial photo analysis. His master's thesis, "Shear Moraines in the Thule Area, Northern Greenland," which was based on this field research, was published as *Research Report 17*, Snow, Ice and Permafrost Research Establishment, Corps of Engineers, U.S. Army, 1957.

While serving in the U.S.Air Force, Bishop was scientific advisor on Rear Admiral Richard E. Byrd 's staff in the U.S. Antarctic Projects Office, Washington, D.C., during 1956-58. He monitored both U.S. and foreign scientific and operations programs and activities, and helped establish and develop a literature exchange program among other nations working on polar research. He was appointed an official U. S. observer with the Argentine Antarctic Expedition in 1956-57.

In May 1959 Bishop joined the editorial and photographic staff of the National Geographic Society as a junior picture editor. A year later he became a Society photographer, eventually moving to the Foreign Editorial staff.

Between May 1960 and January 1961 Bishop participated in the Himalayan Scientific and Mountaineering Expedition led by Sir Edmund Hillary. Under the auspices of the National Geographic Society's Committee for Research and Exploration, Bishop spent six months in the Mingbo Valley, sixteen miles south of Mount Everest, conducting research on the physical geography of the area. On March 13 1961 Bishop scaled the summit of previously unclimbed Mount Ama Dabla (22,494 feet) in the Everest region. From January to June 1963, Bishop took part in the American Mount Everest Expedition and reached the summit of Mount Everest (29,028 feet) on May 22 1963.

Bishop's lectures and writings on the Himalaya fired the imaginations of a whole generation of young Americans. Audiences from those years remember not only his well-organized lectures, his willing counsel, and his encouragement of education through exploration, but also the warmth and good feeling of the social gatherings after the lectures.

In 1964 Bishop was appointed secretary of the Society's Committee for Research and Exploration. In 1966 he took academic leave to pursue a doctorate in geography at the University of Chicago. Between 1968 and 1970 he conducted a dissertation research project on the Karnali Zone in far western Nepal under the auspices of the National Geographic Society, the National Science Foundation, and the Agency for International Development. Under the label of cultural-ecological research, Bishop identified, described, and analyzed the hierarchy of economic systems, including their interaction through the movement of people, animals and goods, and their role in the infrastructure of the larger developing economy of Nepal. He received his Ph.D from the University of Chicago in 1980. Bishop's research was published as *Karnali Under Stress: Livelihood Strategies and Seasonal Rhythms in a Changing Nepal Himalaya*, University of Chicago Geography Research Papers Nos. 228-229, University of Chicago Press, 1990.

During the summers of 1966, 1967 and 1971, Bishop participated in the High Mountain Research Program conducted by the Arctic Institute of North America in Alaska and the Yukon Territory. During a 1966 reconnaissance trip with Dr. Walter Wood of the American Geographical Society, in Saint Elias, Wrangell, and in the Chugach Mountains of the Yukon Territory and Alaska panhandle, Bishop identified the sites which were analogous in selected geo-ecological characteristics to high mountain regions elsewhere in the world.

During the summer of 1967, as research scientist and field coordinator, Bishop established a research station at 17,600 feet on the upper slopes of Mount Logan, and determined the feasibility of conducting physiological and environmental studies at elevations above 14,000 feet. He also participated with Melvin A. Marcus in a research study of the Chitistone Pass area of Alaska.

Bishop's 1971 summer study was aimed at determining the viability of research on the impact of the inhabitants on the alpine environment. This involved in-depth interviews with key people in the Yukon Territory and Alaska, soliciting the views of different ethnic, economic, and political interests toward land use.

Between 1972 and 1977 Barry Bishop served on the editorial staff of the *National Geographic Magazine*. He researched, summarized, and recommended subjects from the broad spectrum of environmental, earth, and social science research for editorial treatment by the Society. In 1980 Bishop became an assistant to President Grosvenor. In 1984 he was named vice chairman of the 17-member Committee for Research and

Exploration and became chairman of the Committee in 1989. In 1992 Barry Bishop became vice president of the National Geographic Society for Research and Exploration.

Bishop's interest and involvement in geographic education coincided with the responsibility for promoting improved communication between the Society and the scientific community. Geographic education in the U.S. went through a period of immense growth and change in the 1980s. As Chief, Geographic Liaison, Office of the President, Bishop put his personal stamp on much that took place in promoting projects of educational value. Barry perceived that to influence the course of geography within the nation, one must be active also within related organizations. Bishop is a member of the American Alpine Club, Explorers Club, International Mountain Society, and a fellow of the Royal Geographical Society. He also serves on the President's Executive Committee of the University of Cincinnati. In 1987 Bishop was awarded The Explorers Club Medal for extraordinary contributions to scientific research.

As vice-president of the National Geographic Society's Committee for Research and Exploration, Barry Bishop continues to strongly foster professional geography.

Richard G. Boehm

Richard G. Boehm
August 29, 1937

Richard G. Boehm, an economic geographer and leader in geographical education, received his Ph.D. from the University of Texas at Austin in 1975. Currently professor and chair of the Department of Geography and Planning at Southwest Texas State University, he has been active in the National Council for Geographic Education, and received the NCGE Distinguished Teaching Achievement Award in 1985. In 1983 Boehm served as president of the NCGE.

Boehm has a longstanding concern for the health and vigor of geographical education. As coordinator of the Texas Alliance for Geographic Education — a network of over 2,800 geography and social studies teachers in Texas — he has raised funds to support summer institutes for teachers from a variety of sources including the state of Texas, National Geographic Society, Southwestern Bell Telephone Company, the Harris and Eliza Kempner Foundation, Sid W. Richardson Foundation, the Strake Foundation, and the Ray C. Fish Foundation. Boehm's summer institutes and workshops have strongly influenced geographic education at the K-12 level in Texas.

Boehm travels widely to conduct workshops and speak to teachers, whom he encourages to opt for graduate education in geography to improve their teaching skills. He played a major role in obtaining a grant from the Fund for the Improvement of Post-Secondary Education for the Association of American Geographers' use in the preparation of model geography courses for teacher training at the university level. He has also prepared materials on environmental education for the U.S. Office of Education.

Boehm is a tireless advocate for the role of geography in public education. He has "presence," boundless energy, and vigor. In the classroom his wide-ranging effective presentations are generally integrated with numerous slides and maps.

One of Boehm's most important attributes is the ability to inspire students, and he gives generously of his time to encourage and guide them professionally. Throughout the American Southwest there are hundreds of teachers who have benefited from his courses and workshops.

As an outstanding teacher, author of geography books and articles, administrator, and director of many educational projects and programs over a period of more than twenty years, Boehm has made a diverse contribution to geographic education.

Harm de Blij

Harm Jan de Blij
October 9, 1935 -

Harm Jan de Blij is one of the most energetic and dynamic members of the geography profession. de Blij was born at Schiedam, the Netherlands, at a time when the rapid pace of world events closed in on space and brought nations closer together. He sensed these global changes and came to represent a revitalized field of geography. de Blij has sought to explain the political geography of the world in a new light. Notwithstanding the attention given these broad visions, he maintains a busy schedule in the practical world of classrooms, committees and study desks, where he constantly works to improve the quality of geography in education.

de Blij received his early education in Europe, and his secondary and college education in South Africa. He graduated from the University of Witwatersrand, Johannesburg, in 1955, with majors in geography and geology. He received his M.A. and Ph.D. in geography from Northwestern University in 1957 and 1959 respectively. After spending a year at the University of Natal as lecturer, de Blij joined the faculty of Michigan State University in 1961, and in 1967 attained the rank of professor. In 1969 he moved to the University of Miami, Florida. In 1980 he was elected to the National Geographic Society's Committee for Research and Exploration, and during 1984-89 was editor of the Society's journal, *National Geographic Research*.

de Blij has conducted research in many areas of the world, but most of his work has been in Africa. Six of his thirty books deal with African topics. During the 1970s, de Blij became interested in the geographical aspects of wine production, consumption, and culture. He visited and studied vineyards and wineries from Japan to New Zealand and from Argentina to Oregon. He published several books on the geography of wine, and wrote numerous articles for scientific and popular magazines. He writes a regular wine column for the American Geographical Society's journal, Focus, and teaches a course on the geography of wine at the University of Miami.

de Blij served as visiting professor at Northwestern and Hawaii, and was Presidential Professor at the Colorado School of Mines. He produced two television series on geography, which aired nationally and in Canada. He is a former editor of the *Journal of Geography* and served as secretary of the Association of American Geographers from 1972 to 1975. He has received numerous grants and awards, including a medal from OIV, the world professional wine organization headquartered in Paris, for his book, *Wine: A Geographic Appreciation.* In 1989, de Blij

Anthony R. de Souza

was made an honorary fellow of the American Geographical Society of New York.

de Blij promotes geographic education through frequent lectures, interviews, and radio and television appearances. He has championed geographic education on the Larry King Show and on ABC Television's Good Morning America program. When he is not writing, editing or teaching, de Blij serves as an enrichment lecturer aboard the cruise ships of the Royal Viking Line.

Anthony R. de Souza
April 23, 1943 -

Anthony R. de Souza is the most widely traveled and experienced geographer of his generation. He contributed greatly to the Association of American Geographers as its secretary during 1986-88, and to the National Council for Geographic Education as the editor of the *Journal of Geography* between 1983 and 1987. In addition he has contributed much counsel as a member of the various committees of the American Association of American Geographers, the National Council for Geographic Education, and the American Cartographic Association. Few others have provided geography with such diversified service.

de Souza was born in Berkhamsted, England, a small town in the Chiltern Hills, in the so-called "cocktail belt," within a short commute of London. A couple of years later his parents moved to Ipswich, the county town of East Suffolk. This historic town with deep roots in the past, some sixty-five miles northeast of London, is a port at the junction of the Sandlings to the east and the rolling glacial clays to the west. It was in Ipswich that young de Souza collected maps, postcards and stamps, and began to dream of faraway places. In his teens he enjoyed traveling throughout East Anglia as a bus conductor during the summers. These early experiences imbued de Souza with the joy of geography which was reinforced by inspiring school teachers.

Between 1962 and 1965 de Souza studied geography at the University of Reading under A. Austin Miller, the climatologist, and beneath a huge portrait of Sir Halford MacKinder, the famous geopolitician and former chancellor of the University. During his undergraduate studies, de Souza developed a strong interest in interpreting landscapes as a whole and learned a good deal about the ingredients — the physical environment, environmental knowledge and perception, cultural ambitions and controls, and technologies used to shape landscapes — needed to understand the changing pattern of human settlement. During this period he also extended his geographic experience through travels in Britain, Ireland, and the Continent.

After graduating with honors from the University of Reading in 1965, de Souza continued to work towards his Ph.D. at Reading, with a concentration in urban-economic geography. His research was on social space and residential mobility in Reading, Berkshire. After completing his field work in 1966, de Souza began writing his dissertation in 1967-68 while on a brief visit to the University of Minnesota. In the summer of 1968 he received his Ph.D. from Reading. At the same time he accepted an invitation from John Webb to be a visiting assistant professor of geography at the University of Minnesota for the 1968-69 academic year.

de Souza, who had developed a strong interest in development geography, left for Tanzania in 1969 to work for the Rockefeller Foundation at the University of Dar es Salaam in both the Department of Geography and the Bureau of Research Assessment and Land Use Planning. Here, between 1969 and 1971, he came into contact with scientists who held a variety of theoretical perspectives varying from free market economics to Third World socialism. Out of this experience, de Souza developed an uneasiness about traditional science and its ability to explain the world. His new thinking is reflected in *The Underdevelopment and Modernization of the Third World,* a Resource Paper published by the Association of American Geographers in 1974. de Souza has used the dialectical approach in this monograph in his teaching at the Universities of Minnesota (1971-75); California, Berkeley (1974); and Wisconsin-Eau Claire (1975-88).

During 1975-87 de Souza devoted much attention to geographic education. He wrote textbooks and published numerous papers on economic geography in peer-reviewed journals, gave workshops and seminars on teaching geography, served as president of the Wisconsin Council for Geographic Education, became contributing editor of the American Geographical Society's *Focus,* and assumed the editorship of the National Council for Geographic Education's *Journal of Geography.*

As editor of the *Journal of Geography,* his overriding goal was to produce a first-rate publication intended to enhance the quality of geographic education and improve geographical literacy at all levels — preschool, through K-12, to undergraduate and graduate study. From 1983 to 1987 the pages of the journal contained resource essays of exceptional depth and breadth written by prominent scholars in American geography. This effort brought new materials and innovative concepts from leading scholars in geography to educators.

In 1987 de Souza joined the staff of the Office of the President of the National Geographic Society in Washington, D.C. to serve as Society liaison with academic institutions and professional societies. In 1988 he was appointed Secretary General of the 27th International Geographical Congress by the U.S. National Academy of Sciences.

In addition to his other responsibilities, de Souza serves as the editor of *National Geographic Research and Exploration*. Under de Souza, *Research and Exploration* has become an intellectually stimulating journal focusing on the natural and human phenomena that make up the world's environments and places. It provides thinking people with a taste of the scope and excitement of geography, and an understanding of the earth as a human habitat. It is this holistic conception of the field which attracted de Souza to geography in the first place.

Robert Gabler

Robert E. Gabler
November 22, 1927

Robert Gabler began his college career in Ohio University, where he majored in geology and received the B.S. degree in 1949. He became interested in geography at Pennsylvania State University, where he received the M.S. in 1951. He began his teaching career in the Stark County, Ohio, public schools in 1950, where he taught for five years before transferring to Columbia University, where he received his doctorate in geographic education in 1957. While living in New York City, he served as lecturer in the Department of Geography at Hunter College and as an instructor in the Teaching of Science Department, Teachers College, Columbia University, during 1955-57.

Gabler returned to the Midwest and joined the faculty at Western Illinois University, Macomb, where he served for six years as an assistant and then associate professor. He became chairman of the Department of Geography and Geology in 1965 and served in that capacity until 1968, when he became chairman of the new Department of Geography. He resigned the chairmanship in 1973. He became director of International Programs at Western Illinois University in 1977, a position he still retains.

In addition to his teaching and administrative duties, Gabler budgeted time for research and writing, publishing four articles in the *Journal of Geography* between 1966 and 1981 and ten articles in other publications. During the same period, he also authored or was multi-author of seven books. The second edition of his introductory college physical geography text was published in 1982.

Gabler is a model teacher in the classroom. He has directed NDEA Teacher Institutes, and has been successful in promoting geography among both geographers and non-geographers. Gabler served as editor and coauthor of *A Handbook for Geography Teachers*, one of the National Council for Geographic Education's major publications. For several years he was editor of the *Bulletin of the Illinois Geographical Society*. He held office in that organization and is a recipient of its Distinguished Geographic Education Award. He has been a member of the Editorial Board of the *Journal of Developing Areas* since 1965.

Gabler has served the National Council well. He began as a member of the Executive Board (1963-68) and Director of Coordinators (1963-68), and rose through the ranks by being elected second vice-president (1972), first vice-president (1973) and president (1974). He played a major role in moving the *Journal of Geography* editorial office to Western Illinois University, and served as business manager of the magazine from January

Charles F. Gritzner

1975 to January 1982. Professor Gabler also played a major role in moving the NCGE's central office to the Western Illinois University campus in June 1980.

As director of International Programs, he administers numerous programs in supervised travel-study. He has also helped design and develop orientation programs for visiting international students.

Professor Gabler has served as a member of: the Illinois Advisory Committee Title III NDEA, (1965-72); the Master Plan, Committee R; the Social Sciences Committee of the Illinois Board of Higher Education (1968-70); the Illinois Task Force on Teacher Certification (1971-73); and the Illinois Task Force on Land Use Planning (a task force convened by the University of Illinois). Professor Gabler has received many recognitions and honors. He was appointed consultant by the U.S. Office of Education for Geography Institute Programs (1965-66). He was invited to give the banquet address at the annual meeting of the NCGE in Des Moines, Iowa, October 1980. He served as representative to the American Association of Colleges for Teacher Education (1971-75). The National Council of Geographic Education selected Robert E. Gabler as the recipient of the George J. Miller Distinguished Service Award in 1980.

Charles F. Gritzner
June 6, 1936

Working quietly and efficiently, Charles F. "Fritz" Gritzner has become a leader in American geographic education. In recognition of his educational leadership, the American Association of Higher Education honored him in 1989. An outstanding teacher, scholar and leader, Fritz has received the Distinguished Service Award from the Association of South Dakota Geographers (1982) and the Distinguished Teaching Achievement Award from the National Council for Geographic Education (1983).

Fritz was born in Fremont, Michigan. As a youngster, he lived near Hess Lake, close to the town of Newaygo. Rural living, environmental awareness, and a penchant for mapping his surroundings in considerable detail engendered in Fritz an early interest in what later became his profession and vocation — geography. In 1949 his family moved to Yuma and from there, a year later, to Mesa, Arizona. Fritz graduated from Mesa

High School in 1954 where he was president of the junior class and of the student body. He was also active in athletics, a factor that influenced his choice of college.

Lured by an athletic scholarship, Gritzner enrolled as a physical education major at Whittier College. After one rather undistinguished semester, he returned to Arizona, enrolling at Arizona State University. It was here during his junior year that Gritzner experienced what he calls "a life changing event." For reasons he no longer recalls, he enrolled in an introductory geography course taught by George T. Renner. Inspired by Renner's lectures, Fritz discovered that his avocation was geography. Until he entered Renner's class, he wanted to be a coach; from that first day in Renner's class he wanted to become a geographer.

Gritzner received his B.A. degree in geography from Arizona State University in 1958, and his M.A. (1960) and Ph.D. (1969) from Louisiana State University. His teaching career began at East Carolina College in 1960. Between 1962 and 1970 he was instructor at Louisiana State University while completing work on his doctoral degree. Fritz credits the profound influence of LSU professors Fred B. Kniffen (folk culture), Robert C. West (Latin America), and anthropologist William G. Haag (concept of culture) with molding his interest in cultural geography. While at LSU, Gritzner also attended several seminars offered by Carl O. Sauer. However, it was H. Jesse Walker, then head of the Department of Geography at LSU, who was instrumental in attracting Fritz to geographic education.

At Walker's initial suggestion, Fritz secured funding for three National Defense Education Act summer geography institutes at LSU in 1965, 1966, and 1968. "The work with teachers was so rewarding," Fritz states, "that I was 'hooked' on geographic education."

During the 1970s he held teaching appointments at East Carolina University (1970-73), the University of Montana (1973-75), Oregon College of Education (1975-76), and the University of Houston (1976-79). In January 1980 he moved to South Dakota State University as professor of geography.

During the last three decades, Gritzner has taught a variety of courses in physical, cultural, and geographic education. He has authored three books, including sixth grade social studies and secondary-level geography textbooks, and prepared the text of a state atlas for students. He has published eleven chapters in edited volumes and more than 125 articles in refereed journals, such as the *Annals of the Association of American Geographers, Journal of Geography, The Professional Geographer, Journal of Cultural Geography, Social Education*, and *The International Journal of Social Education*.

Quite appropriately, Gritzner's first refereed publication was entitled "Let's Sell Geography" (*Journal of Geography,* 1963), which proved prophetic since he has devoted nearly all of his career to doing exactly that — promoting geography.

Gritzner's involvement in and commitment to geographic education has been manifested in numerous ways, including fifteen years as a member of the National Council for Geographic Education Executive Board, serving as executive director (1977-80), and president (1986). He was also chair of the National Council for the Social Studies advisory committee on history, geography and the social sciences (1985-88).

Gritzner has worked closely with teachers throughout most of his career. His office door at South Dakota State University is always open to students, faculty, and others interested in geography.

Gilbert M. Grosvenor

Gilbert Melville Grosvenor
May 5, 1931 -

During the last twenty-five years no American has made such a compelling impact on geographic education as Gilbert M. Grosvenor. Determined to improve the geographic knowledge of Americans — who lag behind their peers in international surveys — the National Geographic Society, under the direction of Gilbert M. Grosvenor, initiated in 1985 a long-term campaign to restore geography education to the nation's classrooms.

Gilbert M. Grosvenor was born in Washington, D.C. and attended Deerfield Academy (1947-50) and Yale University (1950-54), where he recived his B.A. degree in psychology. Although he started out as a pre-medical student, Grosvenor had captured his father's enthusiasm for photojournalism before he finished college. In 1953, during his junior year at Yale, Grosvenor and his roommate read about an international work camp of students forming that summer in the Netherlands, to repair dikes that had been badly damaged by high tides and strong winds. Grosvenor and his roommate signed on, and at his father's suggestion took along their cameras to record the experience. The result was a September 1954 magazine article — and the start of Grosvenor's career at the National Geographic.

After graduating from Yale he joined the staff of the National Geographic Society as a picture editor. He soon left to serve in the U.S. Army's Psychological Warfare Service, resuming his work for the Society and its magazine two years later. Among Grosvenor's various jobs at the Geographic was a two-year apprenticeship in the business department, where he served with the Secretary, the Treasurer, the correspondence office, and the membership-fulfillment section.

Writing and photography assignments for the magazine took Grosvenor around the world, to places like Ceylon (Sri Lanka), Copenhagen, Bali, Yugoslavia, and on President Dwight Eisenhower's diplomatic tour of Asia, Africa and Europe in 1959. Grosvenor's coverage of the tour earned a National Press Photographers Association Award in 1961.

In 1966 he was elected to the Board of Trustees of the Society, and in 1970 was appointed editor of the *National Geographic*. Grosvenor's editorship of the *National Geographic*, from 1970 until he became the Society's 14th President in 1980, has been compared to that of his grandfather. Like Gilbert Hovey Grosvenor (featured in this volume), whose early changes in the magazine were dramatic and reflective of readers' tastes, Gilbert M. steered the magazine into uncharted waters. Under his editorship, it delved into a number of controversial issues, including communism, apartheid, famine, and the horrors of war.

Grosvenor was elected president of the National Geographic Society in 1980 and Chairman of the Board in 1987. He dedicated his career to pursuing a timeless goal, the same as his great-great-grandfather, Gardiner Greene Hubbard, who helped establish the organization in 1888 to increase and diffuse geographic knowledge. That goal, says Grosvenor, is as relevant today as it was then, and it will continue to guide the Society — the world's largest nonprofit scientific and educational institution — into the future. He considers the Geography Education Program created in 1985 as one of the Society's top priorities.

Grosvenor has played a major role in the diversification of the Society, from that of a single-magazine organization to a major publisher of books for adults and children, other magazines, and educational materials. He takes special pride in the success of World, the Society's children's magazine, which started in 1975 and has a circulation of over one million. Grosvenor developed the concept for World — to provide eight-to-thirteen-year-olds with an exciting alternative to watching television.

Together with the staff of the Special Publications Division, Grosvenor also helped create the Books for World Explorers and the pop-up action book series for youngsters. Two new magazines have been launched during his presidency. *National Geographic Traveler*, designed to provide travelers with practical and entertaining travel information, was created in 1984. *National Geographic Research*, a scientific journal that reports the findings of both Society-sponsored and other research, was started in early 1985.

Also in 1985, a decade after the National Geographic Specials debuted on public television, the Society expanded into the world of cable television with the "National Geographic EXPLORER Series," airing weekly on TBS. During the 1980s the Society also greatly expanded its production of educational films and filmstrips for schools. But to Grosvenor, who had grown increasingly concerned about the quality of geography education in the schools, something more was needed.

The Society pledged more than $4 million a year beginning in 1985 to a comprehensive program aimed at improving teacher training and getting students excited about geography. Among its many components is a network of alliances, made up of teachers, administrators, university geographers, and public policymakers working to improve local school curricula and teaching methods. Summer Geography Institutes, held annually at National Geographic headquarters, help teachers increase their geographic knowledge and enhance their teaching of the subject. The National Geography Bee, held annually since 1989, involves 6 million students and awards $50,000 in college scholarships.

Recently, Grosvenor has led the Society into the development of innovative applications of high technology in the classroom, recognizing that "to learn geography in a technological age, our students must have sophisticated technological resources." The Society has created educational programs that incorporate computer software, laser videodiscs and CD-ROM technologies. Working with the National Science Foundation and Technical Research Centers, Inc., of Cambridge, Massachusetts, the Society has established an international information-sharing telecommunications network called Kids Network.

Also during Grosvenor's presidency, the Society opened a new seven-story office building at its headquarters complex in 1984. It houses the 418-seat Gilbert H. Grosvenor Auditorium, which offers extensive educational activities, including illustrated lectures on topics from anthropology to zoology.

Grosvenor is associated with many professional organizations. He was awarded honorary doctoral degrees by George Washington University, the University of Colorado, and the College of Wooster, all in 1983; Curry College in 1984; the College of William and Mary in 1987; and Rhode Island College in 1991.

The International Geographical Congress came to the United States in 1904 under the auspices of the National Geographic — invited by Alexander Graham Bell. In 1992 the 27th International Geographical Congress will convene again in the United States under the auspices of the National Geographic Society and the American geographical community, with Gilbert M. Grosvenor as president of the Organizing Committee.

Grosvenor continues to provide extraordinary leadership to improve geography education at the local, state, and international levels.

Robert A. Harper

Robert A. Harper
April 6, 1924 -

Robert A. Harper, president (1970-71), first vice president (1969-70), second vice president (1968-69), and member of the Executive Board (1967-68) of the National Council for Geographic Education is known to his colleagues as a geographer of broad interests and sympathies, whose ability, modesty, sincerity, patience, and quiet humor enabled him to provide effective leadership in improving geography in America.

Harper grew up in Chicago, Illinois, and entered the University of Chicago on the G.I. Bill after World War II, where he served as an aerial navigator. Harper decided to major in geography to provide a background for a career in journalism. He received a B.S. degree in 1947 and an M.S. in 1948 from the University of Chicago. At Chicago the great influence on his life was Charles Colby, department chair and advisor of his master's thesis. In the beginning, Harper had no idea of pursuing a career in teaching. However, after taking a course on teaching geography from Edith Parker, the geographic education specialist in the department, Harper decided to complete a Ph.D.

C.W. Sorenson, who was associated with Edith Parker in the Barrows-Parker-Sorenson elementary-junior high textbook series, was Harper's classmate. After completing his Ph.D. at Chicago in 1950, Harper joined Southern Illinois University. Sorenson, who was in the geography department at Illinois State University, invited Harper to help him produce geography books for Pakistani schools, for which he had a contract. Harper prepared books on North America, Europe, and on Economic Geography; they were published in Urdu. This marked a fifteen-year association between Sorenson and Harper in writing textbooks. In 1959 Harper wrote a couple of books for Sorenson as part of a revision of the Barrows-Parker-Sorenson series, and in 1962 he joined Sorenson and Raymond E. Crist as joint authors of a junior high social studies text.

Charles Colby, after his retirement from Chicago, came to Southern Illinois University in 1950. Harper and Colby were closely associated for the next fourteen years. Colby, says Harper, was "like a father to me and taught me a great deal about thinking geographically."

In the mid-1960s Harper organized four summer institutes for junior high and high school teachers at Southern Illinois University under the National Defense Education Act, which was part of the post-Sputnik drive to provide a new thrust to the teaching of the school sciences. This led to a number of efforts at improving geography and social studies teaching. Harper became chairman of the geography subcommittee of the High School Social Studies Revision Program in Illinois; was a member of the AAG-NCGE committee which advised Encyclopaedia

Edward P. Hogan

Britannica Films on an ambitious new school media program; wrote a teacher's guide for a course on Latin America for the Greater Cleveland Research Council's Social Studies Program; and was a consultant to the Foreign Policy Association's International Affairs Education Study.

As a result of all this involvement, Harper began writing about geography in general education and social studies education journals. During the 1960s he became active in the Illinois Geographical Society and in the National Council for Geographic Education. He became a member of the Executive Board of the NCGE in 1967 and served as president in 1970-71.

In 1967 Harper moved to the University of Maryland and became professor emeritus after his retirement in 1989. In 1986 he received the George J. Miller Award for Distinguished Service to Geographic Education.

Harper's effective work at Southern Illinois and Maryland in teaching and administration, his counsel to many students, and his contributions to geographical literature have contributed greatly to improving geography in America.

Edward Patrick Hogan
November 12, 1939 -

Edward Patrick Hogan has been of significant service to geographic education. He was born in St. Louis, Missouri, where his father owned and operated a meat processing company. Hogan attended schools in St. Louis, and entered St. Louis University in the fall of 1957.

It was at St. Louis University in 1959, that Hogan, needing a ten o'clock class, took his first course in geography under Professor John Wheedon Conoyer. Conoyer inspired Hogan to pursue a career in geography and also influenced his decision to become an educator. Hogan graduated from St. Louis University in 1961, majoring in geography and business administration. In the fall of 1961 he began working on an M.A. degree in geography at St. Louis University, while teaching in physical geography labs.

After completing his M.A. degree in 1962, Hogan entered the Ph.D. program in sociology-anthropology at St. Louis University. After completing the residency requirements, Hogan began teaching in the Department of Geography as a lecturer. He completed his Ph.D. in 1969 with a combined sociology-anthropology major and geography minor. In his doctoral program he concentrated on demography, social theory, ethnography and area studies, all directly related to geography.

Hogan taught at St. Louis University until 1966. In 1967 he moved from St. Louis to Brookings, South Dakota, as assistant professor of geography at South Dakota State University. This was a new position; while geography had been taught at South Dakota State University when it was established in 1881, geography offerings disappeared only to return briefly during World War II. Hilton M. Briggs, president of South Dakota State University (1958-75), restored geography to the curriculum and invited Hogan to establish the geography program.

In the spring of 1967 Hogan reintroduced geography on the South Dakota State University campus. Through his efforts, an undergraduate geography program was established during the summer of 1967.

Hogan's unyielding enthusiasm has spurred hundreds of classroom teachers to teach geography in new and exciting ways. He worked with grass-roots organizations of university geographers, elementary and secondary classroom teachers, and education decision-makers to elevate the status and effectiveness of geography instruction in South Dakota.

Geography grew rapidly at the South Dakota campus under Hogan's leadership — in the number of majors as well as in total enrollments in geography courses. In 1973 a separate Department of Geography was established with Edward Hogan as head. A Master of Science program in geography was approved in 1974. The Department of Geography also works closely with the Earth Resources Observation System Data Center in Sioux Falls, South Dakota.

In 1976 Hogan was appointed associate dean of the College of Arts and Sciences, in addition to his departmental duties. In 1989 Governor George S. Mickelson appointed him as the State Geographer of South Dakota. Currently, Hogan serves as head of the Department of Geography; associate dean of the College of Arts and Sciences; acting director of the Center for Innovation, Technology and Entrepreneurship; and chair of the South Dakota State University Academic Senate.

Hogan has worked tirelessly for higher quality geography in the schools and colleges in South Dakota. Each year over 3,500 students enroll in it's various programs. About one half of the graduates remain in the state — with the EROS Data Center, state, regional and local government, and private business and industry. The program is also the largest supplier of cartographers for the Defense Mapping Agency Aerospace Center in St. Louis. South Dakota graduates are to be found in geography, cartography, remote sensing, and GIS positions across the country.

Hogan has authored over eight publications. His most recent work is *South Dakota: An Illustrated Geography* (1991), which is being distributed free to some 5,000 high school students. Hogan was instrumental in initiating the South Dakota State Geography Convention in 1970, which is now in it's third decade. The convention annually brings a number of prominent geographers to the campus over a three-day period.

Hogan is dedicated to the importance of excellence in teaching. Even those students at South Dakota who somehow did not have a course with "Dr. Hogan" know of him. He is known as "Ed" to many of the rest of us. Ed Hogan is a man whose influence on geography and geographic education has been great indeed, and whose influence will continue to be felt in the future. There is no finer training for budding undergraduates than to go to South Dakota State University and study under the direction of Ed Hogan. Students receive not only a classic geographic education, but also very strong applied geographic training. Ed is a complete geographer, and his students are well-rounded geographers. For hundreds of South Dakota State University graduates who had the good fortune to sit in on one, or more of Ed's classes, one may rest assured that their lives are being considerably enriched because of this contact.

Lorrin G. Kennamer, Jr.

Lorrin Garfield Kennamer, Jr.
December 20, 1924 -

Lorrin Kennamer, Jr. was born in Abilene, Texas, approximately 140 miles southwest of Fort Worth. Although born in Texas, he spent most of his early life in Kentucky. His undergraduate education was acquired at Eastern Kentucky State College, from which he was awarded the A.B. degree in 1948. For the next two years he taught in the Oak Ridge High School and earned an M.S. degree in geography and education at the University of Tennessee. He continued his student activities at Vanderbilt, George Peabody College for Teachers, and received the Ph.D. in 1952 in geography and education. He then returned to his native state and started his rapid climb to prominence as a geographer and educator.

For four years (1952-56) Professor Kennamer first served as a staff member and later as chairman of the Department of Geography and Geology at East Texas State College. In 1956 he joined the Department of Geography at the University of Texas, becoming chairman of the department and associate dean of the College of Arts and Sciences in 1961. In 1967 he moved to Texas Technological University as dean of Arts and Sciences. In 1970 he returned to the University of Texas as professor of education and dean of the College of Education, a position he held until 1982.

Professor Kennamer is well known for his calm and efficient approach to challenges and changes of college and university life during the past decade. He combines persistence tempered with patience and tenaciousness, softened by a prevailing air of total relaxation. He is recognized by his students and colleagues like as a competent geographer, a thoughtful educator, and a gentleman. His abiding sense of humor makes him a master of ceremonies par excellence — as dean, the president of a professional society, or as a delightful entertaining member of a social gathering (Morris, 1972:504).

Kennamer served as visiting professor during summer sessions at the University of Vermont, Vanderbilt, George Peabody College for Teachers, Michigan State University, and the University of Washington.

Professor Kennamer's contributions went beyond instruction and administration. He served as vice-chairman on the College Entrance Examination Board; president, Southwest Social Science Association (1973); chairman, College Entrance Examination Board (1974-76); chairman of the Commission of Cooperation in Higher Education of the Association of Texas Colleges and Universities; and was a member of the executive Council of the Association of American Geographers. He served as an evaluator of projects for both the Department of Health,

Education and Welfare, and the National Science Foundation. His professional activities included consultant, Bureau of Research, U.S. Office of Education (1966-71); and consultant to the University of Kuwait and the Kuwait Ministry of Education in 1977 and 1978.

Kennamer's bibliography includes seven coauthored books and twenty-one articles; he is contributing author of six additional books. He edited five books and two wall maps, is coeditor of fourteen additional books and manuals, and served as American editor of two books which were published in London. He served as editorial consultant and was a member of the Editorial Advisory Board of the *Journal of Geography* and *Social Education*.

Kennamer's public educational career was interrupted by World War II. He served in the U.S. Naval Reserve, 1943-46. He completed the Navy V-12 program at the University of Notre Dame, attended the Midshipman School of Columbia University, became an Ensign serving as Executive Officer in the LCT Group, Pacific Area, and later served as a lieutenant in occupation duty in Japan.

Because he was such an outstanding student and had a very productive multifaceted career, Kennamer received many honors and awards. After being on the Dean's list for four semesters at Eastern Kentucky State College and at Vanderbilt/Peabody College for Teachers, Kennamer received the Jesse H. Jones Graduate Scholarship (1950-52). He received the Teaching Excellence Award from the University of Texas at Austin (1964); the Distinguished Service Award Vanderbilt/Peabody College for Teachers (1971); the Distinguished Alumni Award, Eastern Kentucky University (1974); was made an honorary life fellow of the Texas Academy of Science; and received the Edward S. Noyes Award for Outstanding service from the College Board of Texas (1982). Also in 1982 the Texas Senate passed a resolution extending warm congratulations to Lorrin Kennamer on receiving the Edward S. Noyes Award and commended him for his outstanding contributions to education.

Although he spent long hours in teaching, research, writing and in administration, Professor Kennamer also served the National Council for Geographic Education in many capacities: secretary, member of the Executive Board, and president (1967). In 1972 he received the Distinguished Service Award of the National Council for Geographic Education.

John F. Lounsbury

John F. Lounsbury
October 26, 1918 -

John F. Lounsbury is recognized as a leader in the field of geographic education who has made an outstanding contribution over many years to improve the geographic knowledge of Americans. For four decades he has contributed to the advancement of geographic education.

Lounsbury was born in Perham, Minnesota. He received a B.S. degree in 1942 from the University of Illinois. After forty months of service in the U.S. Army, including two years as a combat officer in Europe, he began graduate work. He received his M.S. degree from the University of Illinois in 1946 and was awarded his Ph.D. degree from Northwestern University in 1951. While working towards his Ph.D. at Northwestern, he served for two years as the field director of the Rural Land Classification Program of Puerto Rico. This project was initiated by Clarence F. Jones, whose biography also appears in this volume. The undertaking was a basic part of Puerto Rico's well known "Operation Bootstrap" program for developing the island.

In 1951 Lounsbury joined the faculty of Antioch College, where he introduced and developed a geography program. He was appointed chairman of the Department of Earth Science in 1952, where he remained until 1961, when he was appointed head of the Department of Geography and Geology at Eastern Michigan University. The department was developed under the auspices of Mark Jefferson, whose biography is also included in this volume, and was one of the first departments of geography to be established in the United States.

Under the leadership of Lounsbury, the geography program at Eastern Michigan experienced rapid growth and expansion. In 1969 he left Eastern Michigan and assumed the chairmanship of the Department of Geography at Arizona State University, where the department increased in size three-fold and established a Ph.D. program during his eight-year tenure. Following his second four-year term as chair, he was appointed to a three-year term as director of the Center for Environmental Studies. He retired from active teaching in 1987 and is now professor emeritus.

John Lounsbury became involved in educational programs in geography as director of the National Science Foundation Institute a[illegible] Antioch College in 1957. Since then he has directed several nation[illegible] level programs supported by governmental and private foundati[illegible] Perhaps the most widely known program was the Commissi[illegible] College Geography under the auspices of the Association of A[illegible] Geographers, and supported by the National Science Founda[illegible] Lounsbury served as its director from 1965 to 1975. The C[illegible]

ouglas R. McManis

and its working panels included over eighty well-known geographers. The Commission developed and published general papers, resource papers and technical papers; organized institutes and workshops; and developed a consulting service. Overall, the Commission's work had a major impact on instructional and educational programs in American colleges and universities.

John Lounsbury is the coauthor of several college level texts. His *Introduction to Scientific Geographic Research* and *Introduction to Geographic Field Methods and Techniques* reflect his long-time interest in field studies, land classification and land use mapping, and his contribution to the development and application of field techniques.

He served the National Council for Geographic Education as its president in 1979-80, and as a member of several committees. In 1988 John Lounsbury received the National Council for Geographic Education George J. Miller Award for Distinguished Service, and in 1990 he received the Distinguished Mentor Award from the National Council for Geographic Education. His rich experience, breadth of interests, keen insight, and challenging ideas continue to inspire geographers.

Douglas R. McManis

One of the educational goals on which scholars and educators agree is the publication of research results. Students are inspired by faculty members who are excited by their subject and who want their students to share the results of their research. During the last fourteen years Douglas R. McManis — geographer, author and editor — has been bringing to the world community of geographers a balanced survey of research in various areas of the discipline in the *Geographical Review*. It is impossible to determine how many geographers in America and in distant lands have received inspiration from articles in the *Geographical Review*, but it is certain that the editorial work of Douglas McManis has contributed to higher quality geography around the world.

McManis was born in Akron, Ohio, and received his B.S. (1954) and M.S. (1955) degrees from Kent State University. He then went to the University of Chicago for graduate work in geography, and was awarded his Ph.D. in 1964. He taught at the University of Cincinnati (1957-58), Indiana University, Gary Center (1959-60), Butler University (1960-64), and Eastern Michigan University (1964-66). In 1966 he moved to Columbia University and taught there until 1978, when he was named editor of the *Geographical Review*.

G. Etzel Pearcy

His broad knowledge of the field and wide experience makes him equally at home among geographers, ecologists, humanists, land planners, economists and educators, and he is widely respected by all. He has a warm personality and an ingratiating sense of humor. He has contributed to the renaissance in geography through publishing stimulating articles in the *Geographical Review*.

As author and editor, McManis seeks constantly to make what he judges to be the best in geography both interesting and valuable to others. Great ideas can change a discipline, but only if a great editor gives those ideas force. McManis's work is characterized by his intellectual judgment, courage, and dedication to the spread of geographical knowledge. He has the insight, foresight, and willingness to publish bold new ideas.

Surely, Douglas McManis is a geographer who loves this world and the people in it. Authors always consider him as a competent international, national, and local geography expert. Readers of the *Review* share his love for excellence and his real concern for the development of geography. In the geographic profession particularly, throughout America and in many countries overseas, there are hundreds who consider the role of Douglas R. McManis as editor of the *Review* crucial to the renaissance that geographic education is experiencing today.

G. Etzel Pearcy
May 2, 1905 - June 28, 1980

George Etzel Pearcy was one of the most energetic and dynamic of the leaders in geography. He performed significant services to promote geography in business, the government, and among the general public. Etzel Pearcy, more than any other person, made it possible for geographers to use their knowledge and techniques in formulating foreign and defense policies. The little-appreciated and virtually unexploited fields of geopolitics and military geography were recreated by Etzel Pearcy and those who worked with him in the U.S. Department of State and in the Armed Forces.

At no other time in world history had it been so important that major U.S. Government decisions be predicated upon accurate, detailed information about the physical and human environments of the world. Yet, such knowledge was meager at various levels in the federal government. As Geographer in the Department of State, Etzel Pearcy organized the office to provide much of the required data during the post-World War II period. He dedicated his professional career to establishing geography as a significant element in the formulation of American foreign policy. That he was successful is attested by the regard with which the profession is now held, not only in the Department of State, in intelligence agencies, and in broad research programs of the Department of Defense, but in the training of officers for all the services.

Etzel was born in Greencastle, Indiana. He attended Santa Rosa High School, Santa Rosa, California, between 1920 and 1924. He entered the University of California, Los Angeles, in 1924 and received his bachelor's degree in 1930. History, political science, economics, sociology, anthropology, geology, and other fields were all encompassed in this early training. Scientific interest and a sense of curiosity also enriched his background.

He attended Clark University and received an M.A. and a Ph.D., the latter in 1940. He received the American Field Service Fellowship for a year's study at the University of Grenoble in 1933-34. His doctoral dissertation was on the urban geography of Avignon, France. Etzel Pearcy traveled extensively in Europe and Japan in 1935. In the summer of 1936 he traveled in Mexico, Central America, South America, and Western Europe. In 1938 he traveled around-the-world, including New Zealand, Australia, India, France, and Great Britain.

For ten years, between 1933 and 1943, Etzel Pearcy was associated with the academic world. He was instructor in geography and industry at the Wharton School of Finance and Commerce at the University of Pennsylvania (1933-39). In 1939 he was appointed assistant professor of economic geography in the School of Commerce and Business Administration at the University of Alabama. In 1943 he had a temporary war-time appointment as assistant professor at the University of Missouri.

Between 1943 and 1950 Etzel Pearcy was Chief Geographer for Trans World Airlines, Inc. In this position he worked on route development, traffic promotion, and industrial relations for TWA. He helped the Intercontinental Division of TWA develop routes to Mediterranean Europe, North Africa, the Middle East, Western India, and Ceylon following the end of World War II. His research provided the data for TWA to make decisions concerning stops and routes on its international system.

In 1950 Etzel Pearcy joined the Department of State as a Foreign Service Officer. He served as Geographic Attache at the U.S. Embassy in Bangkok, New Delhi, London and Paris with temporary duty in fifty-five other posts between 1950 and 1957. This position involved the procurement of maps and map information for all the official mapping agencies in Washington, including the Library of Congress. It also included negotiations with foreign civilian and military agencies for the exchange of publications and technical information.

Etzel was appointed The Geographer, U.S. Department of State, Washington, D.C., in 1957 and remained there for twelve years. More than any other scholar, G. Etzel Pearcy brought a deep understanding of geopolitics and political geography to Americans after the Second World War. His understanding of the relationships between geographical realities and international affairs was formidable, but more than that, he constantly sought better ways to impact that knowledge and identify the geographical circumstances that most satisfactorily explained the power, interests, character and behavior of nations.

In 1969 he returned to Los Angeles as professor and chair of the Department of Geography at California State University. He retired in 1973 and lived in San Francisco until his death in 1980. During his retirement he continued to write books and papers on a wide variety of topics.

Etzel authored more than forty-five works in political geography and geopolitics. His *Geopolitics in Principle and Practice* (1944) and *World Political Geography* (1948, 1957) were standard college texts for many years. His *Patterns of International Boundaries* (1972), *A 38-State U.S.A.* (1973), and *World Sovereignty* (1977) were widely read. Throughout the United States there are hundreds of teachers who have benefited from his writings. He coedited a series of twenty-five paperback originals on political geography for the D. Van Nostrand Company beginning in 1962.

Etzel Pearcy's style of leadership emphasized innovation in writing and speech as the chief means of winning support for and articulating the dissemination of geography. His books and papers were vehicles for the promotion of geography. Etzel's devotion to geography and its traditions, to his colleagues and geographers around the world, was manifested in many ways. His daily demeanor was one of good cheer.

The University of Kentucky has received funds from Mrs. Etzel Pearcy to establish an endowed academic fellowship in geopolitics in honor of Etzel. The Pearcy Fellowship in Geopolitics will enable graduate students in geography to pursue research on topics such as the geopolitics of peace, neutrality, resource use, international terrorism and human rights, and other related subjects.

Benjamin F. Richason, Jr.

Benjamin Franklin Richason, Jr.
July 24, 1922 - February 9, 1988

Benjamin Franklin Richason, Jr. was born and raised in Loganport, Indiana. He received his undergraduate training at Indiana University, and was awarded the A.B. degree in geography in 1948. Thomas F. Barton became his mentor at Indiana. Barton urged the department to grant Richason a teaching assistantship so he could stay at Indiana University to start his graduate work. Richason was awarded the M.A. in 1949 from Indiana University. With a recommendation from Barton, he became an instructor at Morton Junior College, Morton, Illinois (1949-51).

In 1952 Richason was invited to join Carroll College, Waukesha, Wisconsin, as an assistant professor and chairman of the Geology and Geography Department. He became associate professor (1955-58), then a full professor (1958).

Richason enrolled in the Ph.D. program at the University of Nebraska and received one of the first Danforth Teacher Study Grants in 1955-56. In 1958-59 he received a National Science Foundation grant with which to conduct the research for his dissertation. He received his Ph.D. from the University of Nebraska in 1960.

Although Richason remained chairman at Carroll College for thirty years, he was invited to spend summers and shorter teaching assignments at other universities such as the University of Nebraska, North Carolina Central University, Central Missouri State University, and the University of New Brunswick. He became editor of the *Remote Sensing Quarterly* in 1978.

During his thirty years as department chairman, Richason built a sound, highly respected geography department at Carroll College. Nearly half of his geography majors went on to graduate schools, and fourteen secured their Ph.D.'s.

Although he was a superior teacher and administrator, Richason also had an impressive publication list. He authored or edited four books and monographs, the first of which was *Geography Via Aerial Field Trips.* In 1967 he published an article, "The Audio-Visual-Tutorial Method in Geography," in the *Journal of Geography* and a monograph, *Geography Via the Audio-Visual-Tutorial-Method.* These were followed by *Audio-Visual Tutorial College Units* in 1972. Partly due to these publications, the Audio-Visual-Tutorial Method is now extensively used in some graduate school geography departments. In 1978 he served as editor of *Introduction to Remote Sensing of the Environment* and also published a laboratory manual for the book.

Christopher L. Salter

Professor Richason was an active member of the National Council for Geographic Education, the Association of American Geographers, and Gamma Theta Upsilon. He served as second vice-president, first vice-president, and president (1969) of the National Council for Geographic Education. He was instrumental in publishing the *Geographical Bulletin* of Gamma Theta Upsilon, and was elected president of the international honor society in 1981. He received the Outstanding Science Book Award in geography for 1978 by the Geographical Society of Chicago. In 1976 he was awarded the George J. Miller Award for Distinguished Service to geographic education.

Christopher Lord Salter
August 6, 1938 -

Whether he is teaching a geography class at Columbia, Missouri, leading a group of students on a field trip, or showing school teachers how they can teach geography in new and exciting ways, Christopher L. "Kit" Salter displays a dedication and enthusiasm for his work that inspires those around him. In 1990 the first-ever Distinguished Geography Educator Award of the National Geographic Society went to Salter for extraordinary leadership in increasing geography education in American schools.

Salter was born in Greenport, New York. As a boy, mobility was a way of life for Salter; he attended twenty-two schools before graduating from high school. In the eighth grade he began to hitchhike in an effort to get back to some of the schools that he had left earlier. By the end of his college years at Oberlin College, Salter had hitchhiked more than 125,000 miles. Extensive travel gave Salter an appreciation of small towns, landscapes of fear and uncertainty, and the excitement of coming into an unknown environment. For the young Salter it became an absolute necessity to be able to read these new scenes with some success if he was to get through town after town, urban neighborhood after neighborhood, and curious driver after curious driver with any security or success. Salter's education as a geographer began with these road epics that still glow in his recollections of high school and college years a quarter of a century later.

Salter received his B.A. degree with a major in geology and geography from Oberlin College in 1961. After graduation, Salter and his wife received a two-year teaching assignment at Tunghai University in Taichung, Taiwan, The Republic of China. They had a summer of intensive Mandarin Chinese study at Yale just prior to leaving for Taiwan. On the way across the country to San Francisco to fly to Taiwan, a friend gave Salter a copy of *Asia, East by South* (1954) by Joseph Spencer. It was while reading this book that Salter came to possess — for the first time — any real belief that there was intellectual activity associated with college geography. He said Spencer's book made all the difference in his attitude toward the profession and the discipline of geography.

While in Taiwan, Salter spent considerable time traveling the island by foot, bike, bus and train. As he began to delight in talking to farmers, exploring swidden patches on mountain flanks, and speculating on the impact of Japanese control on the island, he began to see that geography was a uniquely well-suited discipline for accommodating the sort of travel, conversation, and intellectual consideration that he found most engaging. When an oriental librarian from Berkeley came through Tunghai and told Salter of the resources and people in geography at Berkeley, he decided to go there to work towards a doctorate in cultural geography.

Salter spent two years at Berkeley, completing his M.A. in 1967. In 1968 he was invited to join the UCLA faculty and teach East Asian courses. He completed his Ph.D. in 1970. During his nineteen years at UCLA, from 1968 to 1987, Salter moved through the ranks to full professor. Salter's fascination with China was somewhat dampened by the difficulty of field study and travel there. He became interested in the American vernacular landscapes, using the city of Los Angeles as a field laboratory. Salter organized field trips into both American and Mexican landscapes.

In 1984 Salter became involved in an effort to link the UCLA campus with some of the feeder schools that sent students to UCLA. Salter organized afternoon workshops for Grade 7-12 teachers, followed by dinners at the Faculty Club. This dynamic professional interaction became very popular and by early 1985, the California Geographic Alliance was established. The Alliance became the vehicle to bring geography back to the California curriculum, and Salter became involved in these efforts at the state and regional levels.

The California Geographic Alliance created by Salter became the model on which the National Geographic Society's forty-two-state Geographic Alliance Network is based. Salter spent a year (1987-88) as coordinator of the Geographic Alliance Network at the National

Geographic Society in Washington. The Geography Education Program at the National Geographic is the umbrella over the Alliances in forty-two states, the Geography Bee, the nation-wide Geography Awareness Week, and a number of other programs that are having a major impact on American education.

Salter moved to Columbia, Missouri, in October 1988 to take over the chairmanship of the Department of Geography at the University of Missouri. During the past three years, the number of majors at Missouri has grown from twenty-seven to eighty-five. Under the leadership of Salter, the geography program at Missouri is becoming better known for its quality, spirit, and dedication to education.

A prolific researcher and writer in the fields of cultural geography and geographic education, Salter has a far-reaching impact in the classroom. Known as a demanding and dynamic teacher, students and colleagues praise him for his ability to help students grow and develop by generating a curiosity about their subject material and about the world around them.

Joseph P. Stoltman

Joseph P. Stoltman
September 3, 1940 -

Joseph Stoltman is a professor, counselor, and friend to many students at Western Michigan University. His colleagues praise him for his ability to establish an immediate and positive rapport with his students, and he is known for his willingness to spend many hours helping students individually outside the classroom. He has worked tirelessly for more and better geographic education.

Stoltman was born in Seattle, Washington. At the age of four his family moved to Ellensburg in Eastern Washington, where they lived on a cattle ranch located on the Wilson Creek alluvial fan. Stoltman completed elementary, junior high, and high school in the Ellensburg public schools.

Joseph Stoltman's interest in geography was sparked by his sixth-grade teacher, Miss Ellen Munson, who shared an intense interest in ornithology and ecology with her students. Field trips to the flood plain of the Yakima River and bird-observation trips in the vicinity were a regular aspect of geography and science classes. Stoltman's interest in the geography of other countries was sparked by Mrs. Marge Rowley, his seventh-grade geography/language/arts teacher.

His interest in geography was enhanced at Central Washington State College under the tutelage of Professors Robert Funderburk, Martin Kaatz, and Otto Jakubek. Stoltman received his B.A. degree from Central Washington University in 1962. Among the memories of undergraduate education, says Stoltman, were the masterful lectures of Funderburk and the physical geography field trips under the guidance of Kaatz. The Channeled Scablands of Eastern Washington was a regular topic in physical geography and made a lasting impact on Stoltman.

After graduation, Stoltman took a teaching position at Morgan Junior High School in Ellensburg, Washington, while taking graduate courses at Central Washington University and living on the family ranch. His childhood and young adulthood were spent living close to the earth, irrigating fields, harvesting hay and grain crops, and feeding livestock during the winter, which provided Stoltman with a deep appreciation of the environment.

In 1965 Stoltman left rural Washington State and enrolled in the Master of Arts in the Teaching of Geography program at the University of Chicago, where he was introduced to a new way of life in an urban environment and to a geography that he had never imagined in undergraduate years. Gilbert White and others at Chicago provided a new perspective. William D. Pattison supervised his graduate program,

John A. Wolter

and Stoltman received his M.A. from the University of Chicago in 1968.

Stoltman entered the doctoral program in geographic education at the University of Georgia, which was directed by John Ball and Marion Rice, and received his doctoral degree in 1971. The same year he accepted the position of assistant professor of geography at Western Michigan University. He has been active in the International Geographical Union since 1972, a member of the Commission on Geographic Education since 1976, and chair of the Commission since 1980. In working with the Commission, he has encouraged geographers to join an international network for the development and exchange of curriculum and educational research materials.

Stoltman has authored textbooks for elementary and secondary school students, and has published many papers on geography and geographic education in the United States and overseas. He has served as a visiting professor at the University of London and at the University of Zimbabwe. He is currently professor of geography and Distinguished Faculty Scholar at Western Michigan University. He also serves as co-coordinator of the Michigan Geographic Alliance.

In recognition of his numerous services to geographic education, the National Council for Geographic Education honored him with the George J. Miller Award for Distinguished Service in 1989.

John Amadeus Wolter
July 25, 1925 -

John Wolter is a person who has not only attained eminence as a leader and scholar in the world of geographic and cartographic learning, but has also experienced much of the world as an officer in command of large passenger/cargo vessels of the Isthmian Lines on regularly scheduled round-the-world cruises. The author of scores of papers, he puts much energy into enhancing geographic education in America.

When he was very young, he used to collect railroad timetables and maps. He would write for and get maps and timetables from the Southern Pacific, Great Northern, and Southern Railway Systems. He would plot data on his own maps at home. He read all the travel books he could get his hands on, and was fascinated with "just the idea of seeing things."

From the youngster who dreamed of seeing exotic places to the merchant seaman who actually saw those places, John Wolter became the principal geographer in the Library of Congress. The transition from adventurous merchant seaman to chief of the Geography and Map Division may seem a peculiar one, but the roster of places he has visited and the experiences he has had are enough to make any explorer envious.

A native of St. Paul, John Wolter is a graduate of the University of Minnesota (B.A. in geography, 1956), from which he received a master's degree in library science (1965) and a Ph.D. in geography (1975). He was map librarian at the University of Minnesota Library (1961-64 and 1966) and a lecturer in the Department of Geography during the same period. He served as assistant to the Director of Libraries, University of Minnesota, in 1964-65; was an assistant professor in the Department of Geography at the University of Wisconsin, River Falls, from 1966 to 1968, and served as acting chair during 1968.

John Wolter joined the Library of Congress as assistant chief of the Geography and Map Division in August 1968. He was promoted to chief of the Geography and Map Division in 1978. In 1989 he became acting director of Public Services and Collections Management. In this position he provided coordination and leadership to eight area and special collections divisions (Geography and Map Division, Manuscripts, Music, Prints, Photographs and Motion Pictures, Rare Books, Africa and the Middle East, and Asian). He served as a principal spokesperson for the Library on issues related to public service and collections development. Wolter retired in August 1991.

As chief of the Geography and Map Division between 1978 and 1991, he served as the senior information specialist on cartography and geography in the Legislative Branch of the U.S. Government. He was responsible for the acquisition, classification, custody, preservation, and effective use of the largest and most comprehensive collection in the Western Hemisphere, which includes over four million maps, charts, atlases, globes, models, and other forms of cartographic materials.

As the principal geographer in the Library, John Wolter had the professional responsibility for the evolution and direction of research and development programs in the fields of geography and cartography.

Active in professional societies and organizations, John Wolter served some fourteen years as the Library's member of the U.S. Board on Geographic Names and as Board Chairman (1980-83). He was also chairman of the Board's Centennial Celebration Committee in 1990. He served as a government representative to the U.S. National Committee for the International Geographical Union during 1972-80 and 1984-88. He is a past president of the Society for the History of Discoveries and a director of the New Mexico Geographical Society.

By all accounts, John Wolter is a leader with candor, imagination, and pride.

References

Ackerman, E. 1967. Gilbert Hovey Grosvenor, 1875-1966. *Annals of the Association of American Geographers.* 56:598-599.

Anderson, H. 1950. Offerings and Registrations in the Social Studies. *Social Education* 14:73-75.

Anderzhon, M. 1979. In Memoriam: Mary Viola Phillips. *Journal of Geography.* 78:6.

Bacon, H. 1955. Fireworks in the Classroom: Nathaniel Southgate Shaler. *Journal of Geography.* 54:349-353.

Barton, T. 1973. George J. Miller, February 8, 1880-July 23, 1973. *Journal of Geography.* 73:53-59.

Barton, T. 1971. Mamie Anderzhon Receives the 1971 Distinguished Service Award. *Journal of Geography.* 70:562-563.

Barton, T. 1969. Distinguished Service Award. *Journal of Geography.* 68:560.

Barton, T. 1965. Publications of Leaders in Geographic Education, Report II, Katheryne Thomas Whittemore. *Journal of Geography.* 64:211-213.

Barton, T. 1964. Leadership in the Early Years of the National Council of Geography Teachers, 1916-1935. *The Journal of Geography.* 63:345-355.

Barton, T. 1950. Edith Parker Receives the Distinguished Service Award in Geographic Education. *Journal of Geography.* 49:49-50.

Bauleg, H. 1950. William Morris Davis: Master of Method. *Annals of the Association of American Geographers.* 40:188-195.

Black, W. 1986. Thomas Frank Barton, 1905-1985. *Journal of Geography.* 85:42-43.

Bowman, I. 1950. Mark Jefferson. *The Geographical Review* 40:134-137.

Bowman, I. 1934. William Morris Davis. *The Geographical Review.* 24:177-181.

Brigham, A. 1927. Contribution of Geography in Senior High Schools. *The Journal of Geography.* 26:165-174.

Brown, R. 1941. The American Geography of Jedediah Morse. *Annals of the Association of American Geographers* 31:145-217.

Bryan, K. 1935. William Morris Davis — Leader in Geomorphology and Geography. *Annals of the Association of American Geographers.* 25:23-31.

Burnham, G. 1953. Douglas C. Ridgley, 1868-1952. *Journal of Geography.* 52:82.

Carter, G. Isaiah Bowman, 1878-1950. *Annals of the Association of American Geographers.* 40:335-350.

Chamberlain, J. 1939. Early American Geographies. *Yearbook of the Association of Pacific Coast Geographers.* 5:23.

Clark, R. 1934. *Geography in the Schools of Europe in Relation to a Program of Geographic Instruction in American Education.* New York City: Charles Scribners Sons.

Colby, C. 1960. Dr. Vernor C. Finch, 1883-1979. *Journal of Geography.* 59:239.

Colby, C. 1933. Ellen Churchill Semple. *Annals of the Association of American Geographers.* 23:229-240.

Commager, Henry Steele. 1950, *The American Mind.* New Haven: Yale University Press.

Cressey, G. 1949. Wallace W. Atwood. *Annals of the Association of American Geographers.* 39:296-306.

Cunningham, F. 1967. Distinguished Service Award. *Journal of Geography.* 66:522.

Cunningham, F. 1965. Distinguished Service Award. *Journal of Geography.* 64:423.

Davis, W. 1924. The Progress of Geography in the United States. *Annals of the Association of American Geographers.* 14:159-214.

Davis, W. 1902. The Progress of Geography in the Schools. In *First Yearbook of the National Society for the Study of Scientific Education Part II,* ed. M. Holmes, pp. 7-47. Chicago: University of Chicago Press.

Dodge, R. 1934. William Morris Davis — An Appreciation. *Journal of Geography.* 33:148-150.

Dryer, C. 1924. A Century of Geographic Education in the United States. *Annals of the Association of American Geographers.* 14:117-149.

Entrikin, N. and Brunn, S. 1989. *Reflections on Richard Hartshorne's*

The Nature of Geography. Washington, D.C: Association of American Geographers.

Fairbanks, H. 1927. *Real Geography and Its Place in the Schools.* Wagner: San Francisco.

Finch, V. 1939. Ray Hughes Whitbeck, 1871-1939. *Journal of Geography* 28:252-

Foscue, E. 1936. The Place of Geography in the Senior High School with Special Referemce to Texas. *The Journal of Geography* 35:117-122.

Foster, A. 1955. The New Department and Its Setting. *A Half Century of Geography — What Next?* Department of Geography, University of Chicago. pp. 5.

Foster, A. 1937. The Distinguished Service Award. *Journal of Geography.* 36:71-72.

Freeman, A. 1954. Eerna Grassmuck Gilland Receives Distinguished Service Award — National Council of Geography Teachers. *Journal of Geography.* 53:364-365.

Gabler, R. 1982. In Memoriam Herbert Henry Gross, 1905-1981. *Journal of Geography.* 81:78-79.

Griffin, P. 1952. Richard Elwood Dodge: Pioneer in Educational Geography. *Journal of Geography.* 51:253-258.

Gritzner, C. 1977. George J. Miller for Distinguished Service. *Journal of Geography.* 76:276.

Haas, W. and Ward, H. 1933. J. Paul Goode. *Annals of the Association of American Geographers.* 23:241-246.

Harper, R. 1966. Obituary, Charles Carlyle Colby, 1884-1965. *The Geographical Review.* 56:296-297.

Harris, C. 1968. Stephen Sargent Visher, 1882-1967. *Journal of Geogrpahy.* 67:378-379.

Harris, C. 1966. Charles C. Colby, 1884-1965. *Annals of the Association of American Geographers.* 56:378-382.

Harris, C. 1963. Alice Foster, 1872-1962. *Journal of Geography.* 62:77.

Hartshorne, R. 1964. Robert S. Platt, 1891-1964. *Annals of the Association of American Geographers.* 54:630-637.

Hartshorne, R. 1961. Vernor Clifford Finch. *Annals of the Association of American Geographers.* 51:339-342.

Herman, T. 1965. George Babcock Cressey, 1896-1963. *Annals of the Association of American Geographers.* 55:360-364.

Hewes, L. 1964. Nels August Bengtson, 1879-1963. *Journal of Geography.* 63:85.

Hewes, L. (n.d.) *Dr. Nels A. Bengtson Memorial Scholarship in Tribute to a Geographer-Geologist.* Lincoln: University of Nebraska Foundation.

Houk, R. 1973. Distinguished Service Award. *Journal of Geography.* 72:61-62.

Houk, R. 1990. Elizabeth Eiselen. *Journal of Geography.* 89:113.

James, P. 1974. Distinguished Service Award. *Journal of Geography.* 73:62-63.

James, P. 1964. George Babcock Cressey, 1896-1963. *Geographical Review.* 54:254-257.

Kemter, J. 1962. Edith Putnam Parker, 1886-1961. *Journal of Geography.* 61:222.

Koelsch, W. 1979. Nathaniel Southgate Shaler, 1841-1906. *Geographers Bibliographic Studies.* 3:133-139.

Leighly, J. 1978. Carl Ortwin Sauer, 1889-1975. *Geographers Biographical Studies.* 2:99-108.

Leighly, J. 1956. R. D. Calkins, 1873-1955. *Journal of Geography.* 55:43.

Marshall, H. 1956. *Grandest of Enterprises, Illinois State Normal University, 1857-1957.* Chicago: Lakeside Press, R.R. Donnelley and Sons.

Manson, G. 1981. Notes on the Status of Geography in American Schools. *The Journal of Geography.* 81:244-248.

Martin, H. 1964. Otis Willard Freeman, 1889-1964. *The Journal of Geography.* 63:332.

Martin, L. 1950. William Morris Davis, Investigator, Teacher and Leader in Geomorphology. *Annals of the Association of American GeographersP.* 40:175-180.

McMillion, O. 1952. Geography in the Teachers Colleges. *Professional Geographer.* 4:17-19.

Meyer, A. 1959. Citation: Distinguished Service Award to Katheryne Thomas Whittemore. *Journal of Geography.* 58:458-459.

Meyer, A. 1946. College Geography and Its Relation to Teacher Training in Secondary School Geography. *The Journal of Geography* 45:45-78.

Miller, R. 1948. High School Geography in Nebraska. *The Journal of Geography* 47:8-17.

Miller, T. 1966. A History of the Geography Department of Valley City State College. *Bulletin of the Association of North Dakota Geographers.* April, pp. 15-23.

Morris, J. 1972. Distinguished Service Award. *Journal of Geography.* 71:504-505.

Morris, J. 1968. Joe Russell Whitaker, Teacher-Geographer. *The Peabody Reflector.* :189-191.

Orchard, D. 1967. Obituary. *The Geographical Review* 57:128-130.

Pearcy, E. 1958. George Renner, 1900-1955. *Annals of the Association of American Geographers.* 48:244-249.

Phelps, J. 1970. Distinguished Service Award. *Journal of Geography.* 69:558.

Phelps, J. 1962. M. Melvina Svec Receives Distinguished Service Award, National Council for Geographic Education, 1962. *Journal of Geography.* 61:418.

Phillips, M. 1965. Zoe Agnes Thralls, 1888-1965. *Journal of Geography.* 64:299.

Phillips, M. 1964. *Edith Putnam Parker: Her Work and Contributions to Geographic Education.* Ann Arbor: University of Michigan.

Platt, R. 1950. Isiah Bowman, 1878-1949. *Journal of Geography.* 49:174.

Powell, P. 1911. General Science versus Physical Geography. *Social Science and Mathematics.* 11:119.

Read, M. 1949. Stephen S. Visher Receives the Distinguished Service Award in Geographic Education. *Journal of Geography* 48:41.

Rose, J. 1971. Stephen Sargent Visher, 1887-1967. *Annals of the Association of American Geographers.* 61:394:406.

Rose, J. 1964. Griffith Taylor, 1880-1963. *Annals of the Association of American Geographers.* 54:622-629.

Rosen, S. 1947. A Short History of High School Geography. *The Journal of Geography* 46:405-411.

Royan, W.Van. 1968. Nels August Bengtson, 1879-1963. *Annals of the Association of American Geographers.* 58:602.

Schwendeman, J. 1956. *Directory of the College Geography in the United States.* 9.

Schwendeman, J. 1952. *Directory of the College Geography in the United States.* 5.

Shrode, I. 1929. A Catalogue Study of Geography in Educational Institutions Above the High School. *The Journal of Geography* 28:189-201.

Smith, G. 1959. Armin Kohl Lobeck, Geomorphologist and Landscape Artist, 1886-1958. *Annals of the Association of American Geographers.* 49:83-87.

Stanislawski, D. 1975. Carl Ortwin Sauer, 1889-1975. *Journal of Geography.* 74:548-554.

Starkey, O. 1967. Joseph Russell Smith, 1874-1966. *Annals of the Association of American Geographers.* 57:198-202.

Svec, M. 1968. Distinguished Service Award. *Journal of Geography.* 67:504.

Svec, M. 1960. Earl Bennett Shaw Receives Distinguished Service Award. *Journal of Geography.* 59:434.

Tuthill, R. 1956. George Thomas Renner, 1900-1955. *Journal of Geography.* 55:93-94.

Uttley, M. 1964. Alison Esther Aitchison, 1874-1964. *Journal of Geography.* 63:278.

Valkenburg, S. 1968. The Making of a Geographer. *Journal of Geography.* 67:524-525.

Valkenburg, S. 1948. Ellsworth Huntington. *Geographical Review.* 38:153-155.

Villmow, J. 1976. George J. Miller Award for Distinguished Service. *Journal of Geography.* 75:575-576.

Visher, S. and Hu, C. 1950. Oliver Edwin Baker, 1881-1949. *Annals of the Association of American Geographers.* 40:328-336.

Visher, S. 1953. Rollin D. Salisbury and Geography. *Annals of the Association of American Geographers.* 42:4-11.

Visher, S. 1949. Mark Jefferson, 1863-1949. *Annals of the Association of American Geographers*. 39:307-312.

Visher, S. 1947. Ellsworth Huntington, 1876-1947. *Science*. 106:632-633.

Warman, H. 1975. The George J. Miller Award for Distinguished Service. *Journal of Geography*. 74:508-512.

Warman, H. 1949. Dr. Wallace W. Atwood, 1872-1949. *Journal of Geography*. 48:350.

Wetmore, A. 1966. Gilbert Hovey Grosvenor, 1875-1966. *Geographical Review*. 56:598-599.

Whipple, G. 1933. *The Teaching of Geography, Thirty-Second Yearbook of the National Society for the Study of Education*. Bloomington, Ill: Public School Publishing Company.

Whitaker, J. 1941. Almon Ernest Parkins, 1874-1940. *Annals of the Association of American Geographers*. 31:46-50.

Whitaker, J. 1940. Almon Ernest Parkins, 1879-1940. *Journal of Geography*. 39:167-168.

William, F. Ray Hughes Whitbeck: Geographer, Teacher and Man. *Annals of the Association of American Geographers* 30:211-217.

Winstead, H. 1912. Geography in American Universities. *The Journal of Geography*. 10:309-316.

Wright, J. 1966. A Plea for the History of Geography. *Human Nature in Geography*. Cambridge: Harvard University Press.

Wright, J. 1959. Some British "Grandfathers" of American Geography. In *Geographical Essays: A Memorial Volume for the late Alan G. Ogilvie of Edinburgh University*, eds. R. Miller and J. Wredford, London: Longmans Green Company.